Concurrent Engineering

M. Adithan

ATLANTIC
PUBLISHERS & DISTRIBUTORS (P) LTD

Published by

ATLANTIC

PUBLISHERS & DISTRIBUTORS (P) LTD

7/22, Ansari Road, Darya Ganj, New Delhi-110002
Phones : +91-11-40775252, 40775214, 23273880, 23275880
Fax : +91-11-23285873
Web : www.atlanticbooks.com
E-mail : orders@atlanticbooks.com

Printed & bound in India by Atlantic Print Services

Dedicated to

Today's Students, and

Tomorrow's Engineers

Preface

Concurrent Engineering (CE) is relatively a new concept in manufacturing and technology, and is fast catching up with the industries. Indian industries which have implemented CE in their design, development, and manufacturing activities have reaped many benefits such as reduced manufacturing lead time, have become agile, and are able to incorporate quick design changes and variations in a short-time and continue to remain competitive.

Realizing the significance of this subject, many universities have incorporated "Concurrent Engineering" as a full-fledged course in their mechanical engineering curriculum. Students of Manufacturing, Production, and Mechanical Engineering in general should have a good knowledge of this subject which will be useful to them when they enter the world of work and start building up a career in engineering/mechanical engineering.

With this in view, this textbook has been developed to give an impetus to this new and innovative concept.

M. Adithan

Acknowledgements

I would like to acknowledge and thank my colleagues and authors of various conference papers presented and books published which enabled me to gain further insights into the topic of "Concurrent Engineering".

- Robert C. Creese, L. Ted Moore, and J. Cleetus, West Virginia University, USA, 2010.
- Society of Automotive Engineers, USA, 1986.
- Society of Manufacturing Engineers, USA, 1985.
- Mikell P. Groover, USA, 1992.
- Devi Thirupathi, Samundeeswari, and S. Sarukesi Karunakaran, 1999.
- R. Sivasubramanian, V. Selladurai, and N. Rajanramasamy, 1999.
- V. Ananda Kumar, V.K. Sethi, and S.C. Chetal, 1999.
- David D. Bedworth, Mark R. Henderson, and Philip M. Wolfe, USA.
- Arthur L. Foston, Carolena L. Smith, and Tony Au, 1991, USA.
- Harold T. Amrine, John A. Ritchey, Oliver S. Hulles, Colin L. Moodie, and Joseph F. Kmes, 1982, USA.
- K.P. Karunakaran, Vijay P. Bapat, Srenath Babu Akula, and P.D. Solanki, 2001.
- Trevor J. Bentley, UK, 1980.
- Thomas Pyzdek and Roger W. Berger, 1991, USA.
- Shubha Jain, 2010.
- David Valliere, USA, 1990.
- George E. Dieter, USA, 1991.

- Don Weller and Michael Clements, Geneva, Switzerland, 1979.
- Daniel T. Koenig, USA, 1986.

I appreciate the outstanding contribution made by the above authors for the cause of manufacturing engineering and technology education which inspired me to develop this text. I appreciate their scholarship, R&D efforts, dedication, and commitment to manufacturing engineering and technology education in India and abroad.

M. Adithan

University Syllabus

Mechanical Engineering

University Elective: Concurrent Engineering

Unit I

1. Introduction to Concurrent Engineering
2. Use of Information Technology

Unit II

3. Design Stage in the Life Cycle of Products

Unit III

4. Manufacturing Concepts and Analysis
5. JIT System
6. Design of Automated Manufacturing System

Unit IV

7. Product Life Cycle and DFM (Design for Manufacturing)
8. Manufacturing Cost

Unit V

9. Concurrent Mechanical Design
10. Plan for Project Management

Contents

Preface .... *v*
Acknowledgements .... *vi*
University Syllabus .... *viii*

1. Introduction to Concurrent Engineering (CE) 1
Different Definitions of Concurrent Engineering
Benefits of Concurrent Engineering
Collaborative Product Development
Organizing for Concurrent Engineering
Concurrent Engineering: Tool Box (Techniques)
Evaluative Questions

2. Use of Information Technology (IT) 13
Information Technology Support
Solid Modelling
Product Data Management (PDM)
Collaborative Product Commerce
Artificial Intelligence (AI)
Expert System (ES)
Software and Hardware Co-design
Evaluative Questions

3. Design Stage in the Life Cycle of Products 29
Product Life Cycle
Modality of Concurrent Engineering Design
Concurrent Engineering in Optimal Design
Opportunities for Manufacturing Enterprises
Real Time Constraints
Evaluative Questions

4. Manufacturing Concepts and Analysis 41
Manufacturing Competitiveness
Conceptual Design Mechanism
Checking the Design Process
Manufacturability/Producibility
Intelligent Design for Manufacturing System
Evaluative Questions

5. JIT System 54
What is Inventory?
Importance Maintaining Low Inventory
Modular Design/Kitting
Evaluative Questions

6. Design of Automated Manufacturing System 60
Automation and Mechanization
Stages in Development of Automated System
Advantages of Automation
Assembly Planning
Evaluative Questions

7. Product Life Cycle and Design for Manufacturing 70
Design for Economics
Design for Assembly
Evaluative Questions

8. Manufacturing Cost 77
Economical Lot Size
Fixed Cost,Variable Cost and Overhead Charges
Break-even Analysis
Evaluation of DFM
Evaluative Questions

9. Concurrent Mechanical Design and Product Realization 90
Geometric Modelling and Primitives
Decomposition in Concurrent Design
Product Realization: Taxonomy
(Different Terms Associated with "Product Design")

Negotiation in Concurrent Engineering Design Studies
Evaluative Questions

10. Plan for Project Management on New Product Development 100
Planning a Design Project
Stages in Project Management/Steps
CPM and PERT
Bottleneck Technology Development
Evaluative Questions

Glossary of Terms Related to Concurrent Engineering 107

University Model Question Paper 137

Index 142

1

Introduction to Concurrent Engineering

Concurrent Engineering is a systematic approach to the integrated, concurrent design of products and their related processes: manufacturing and support services. This approach helps the product designers and industrial designers, in the initial stages itself to consider all elements of product life cycle, including quality, cost, time schedule (time frame) and user requirements.

Different meanings of Concurrent Engg: There are many *names and terms associated with concurrent engineering*. These are:

- Simultaneous Engineering (in contrast to sequential approach)
- (Computer) Integrated Design and Manufacturing
- Horizontally integrated manufacturing
- Integrated cross-functional engineering
- Design for manufacturing (DFM), Design for Testing, Design for Assembly (DFA): [Design for manufacturing and assembly, DFMA]
- Producibility Engineering or Manufacturability (involving integrated consideration of Design, Manufacturing and Material).
- Holistic approach to design and manufacturing.
- Collaborative Engineering or Collaborative Manufacturing
- Internet based/web-enabled design and manufacturing

- e-manufacturing
- Synchronous manufacturing
- Systems Engineering

Concurrent Engineering

There are two important engineering functions carried out in an engineering industry (manufacturing industry):

(i) Component design (or) *Product Design*

(ii) Manufacturing of the part/product (sometimes called *process design*)

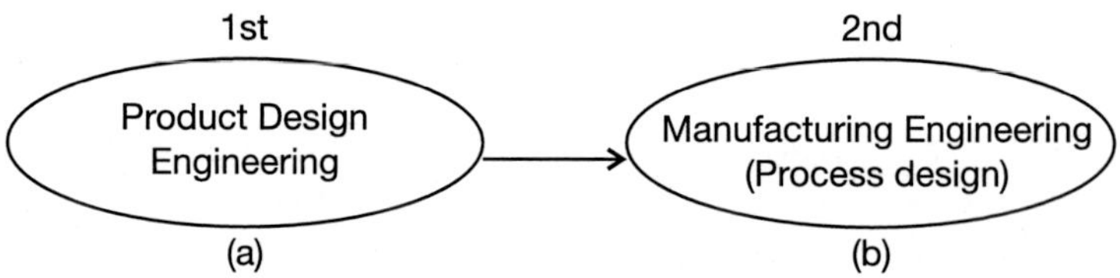

Fig. 1.1: Traditional Approach—(a) Product design and (b) Process design

This is shown in Fig. 1.1. Traditionally these two functions are carried out at two different departments in an industry, or in some cases, at two different locations/regions, separated by organizational walls and distance.

The product design details and requirements such as drawings, tolerances and materials are communicated to manufacturing department for their consideration. If there is any change is required, considering the manufacturing capability, the same is communicated or sent to product design dept. one after another for design, if required. Thus *information/communication is sent to and fro* involving time delay. This is avoided in concurrent engineering since all aspects, especially design and manufacturing are considered by the *cross-functional unit* formed in *an integrated manner concurrently*. This is considered as *an unique* feature of concurrent engineering, and many advantages and benefits accrue from this concept and approach (Fig. 1.2).

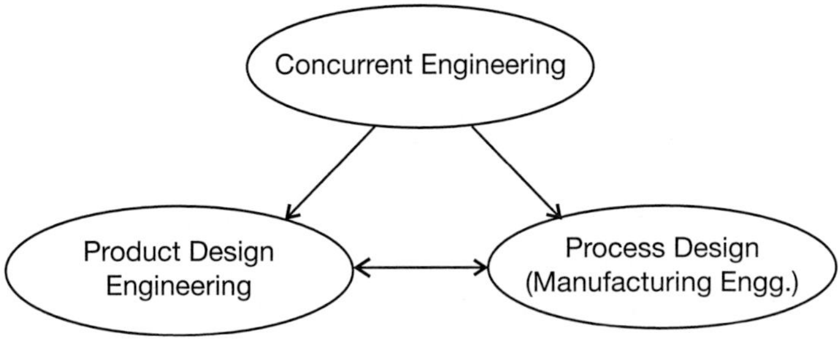

Fig. 1.2: Concurrent Engineering approach integrating design and manufacturing functions simultaneously and in the initial stages itself. Concurrent engineering is a design approach in which product design and product manufacturing are merged in an intimate way

> Concurrent engineering is a management and engineering philosophy aims to improve quality, decrease costs, and reduce the lead time from concept to through development of new products.

Different definitions of Concurrent Engineering: There is no single definition of concurrent engineering.

- *Concurrent Engineering* is the process of designing a product using all inputs (information) and evaluations *simultaneously* and early during design stage itself to assure that both internal customers (different departments) and external customers (users of end product) needs are met.
- *Concurrent Engineering*, also known *as collaborative engineering*, is the systematic approach of doing all manufacturing related activities *at the same time* in an *integrated manner*, as far as possible, and preferably at the *design stage itself*. Concurrent engineering minimizes modifications and changes in product design/part design at a later stage.
- *Concurrent Engineering* (*or Simultaneous Engineering*) "Concurrent Engineering" is also referred to as "simultaneous engineering." Both 'concurrent' and 'simultaneous' may have the same meaning: "to *happen at the same time*", but there is *a subtle difference*:

the word "concurrent" has the additional meaning "*agreeing*", "*co-operating*". The purpose of both (concurrent engineering and simultaneous engineering) is to deliver the product/carry-out the activities in the shortest time possible and at the highest quality level and it necessitates the focussing of the total efforts towards achieving a common goal.

- *Concurrent Engineering* integrates synergitically technological processes, people (team; engineers and managers) and information systems to enable a task to be performed effectively and timely (Fig. 1.3).

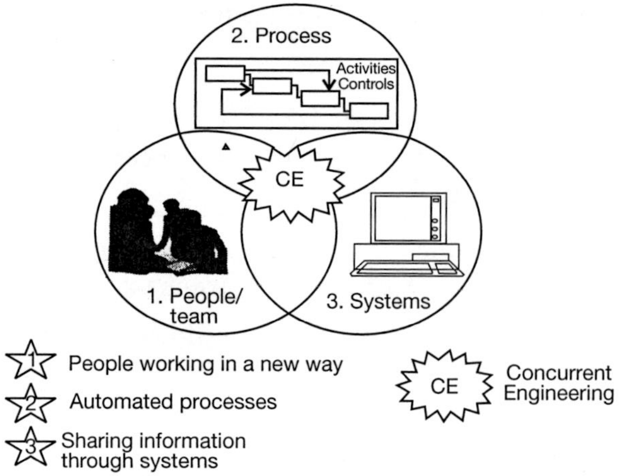

Fig. 1.3: A structure of concurrent engineering (Brooks 1995) Concurrent Engineering integrates the processes, People (Team) and the systems simultaneously performing the task effectively

Concurrent Engineering
Concurrent Engineering is a *systematic approach to integrated product design and development meeting customer requirements* with co-operation, trust and *sharing of information amongst people and teams* and by working *simultaneously* and *collaboratively*.
–Cleetus, 1992

An important aspect of concurrent engineering is to incorporate the idea *designing for lifetime of use, i.e. life cycle engineering.*

According to Institute for Defence Analysis, USA

Concurrent Engineering is a systematic approach to the integrated, concurrent design of products and processes, and manufacturing from the outset, to consider all the elements of the product cycle from conception to the disposal of the product.

The aim of concurrent engineering is to reduce the time from product concept to market, prevent quality and reliability problems, and reduce costs.

Traditionally, activities during product development are handled *segmentally*, not concurrently, common practice is marketing or research department identifies a product idea (new product); next, design engineering develops a design and builds a few prototype (trial) units; then the purchasing department calls for quotations (bids) from vendors/suppliers, after that manufacturing department produces the units. At each step, the output information from one department is transferred to the next department. Design changes or any other change information is also done in this manner. This results in time delay in executing the modifications (Fig. 1.4a). This approach is called sequential or traditional approach. Predominantly *uni-directional information flow* takes place.

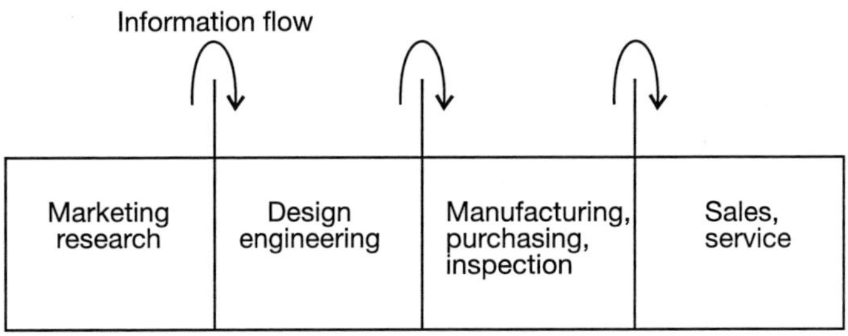

(a) Sequential engineering

Fig. 1.4a: Information/design and design changes are transferred from one dept. to another dept. in a *sequential* way, involving time delay.

In Concurrent Engineering, all the design parameters: i.e. basic functional requirements, reliability, maintainability, safety, human factors, manufacturability, inspection, packaging, transportation and storage are addressed, concurrently, i.e. simultaneously resulting in many benefits to the company (Fig. 1.4b)

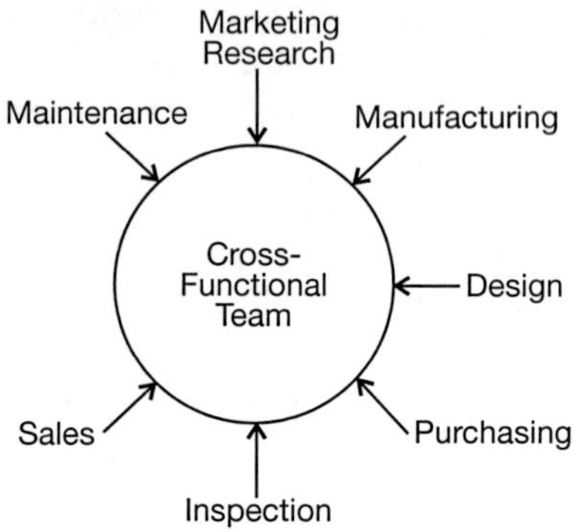

(b) Concurrent Engineering

Fig. 1.4b: In concurrent engineering design changes made are accessible to all functional groups (departments) *concurrently* or *simultaneously* for corrective action quickly, through formation of cross-functional team.

In concurrent engineering, departments have early access to design reviews/design change information. Future problems, likely to arise can be identified and avoided.

In concurrent engineering *multi-directional information* flow takes place, i.e. inputs from other departments and functional groups are taken.

Benefits of Concurrent Engineering (based on actual industry case studies/examples), as reported in publications:

- 75% fewer engineering design changes (since design validation is done in the initial stages)
- 55% less time from product concept to market (time to market)
- Improved design quality
- Quality and reliability problems are reduced
- Reduction in overall manufacturing cost (30%-40%) (less time over-run and cost over-run)

- The development time from product conception to market (i.e. time to market) is reduced since activities are done in parallel instead of in series
- Manufacturing lead time is also reduced by about 30%
- Synergistic effects of integrated design and manufacturing results in less cost
- Scrap and rework reduced by as much as 75%
- Effective communication and interaction amongst different departments results in better coordination and greater understanding of the product requirements and how to achieve it.

All the above benefits improve the manufacturing competitiveness of the industry.

Manufacturing Lead Time (MLT)

It is the total time required to process a work order. It is the *time between customer order and product delivery*. Less manufacturing lead time gives good customer satisfaction. MLT depends on no. of operations, set-up time and time/ operation (cycle time) and batch size if the order is for a batch size, usually expressed in weeks.

As a result of introduction of CE (concurrent engineering) an important advantage claimed by industries is, *reduction in manufacturing lead time* by as much as 30%.

In concurrent engineering, "*disposability*" (*salvage*) of the used part is also considered at the design stage itself. This is a new dimension in CE in view of pollution/environmental concerns/issues.

Product Planning

Product planning is an activity that involves:

- making certain that new products are added,
- old products are discontinued, and
- certain products, changed at the appropriate time and continues to be produced and marketed with changes in design and in new product ranges.

To enable this activity to be performed efficiently clear-cut objectives and policy guidelines should be established by the company. Adequate funding/budgetary allocation should be made for new product development. Sales forecasting and market studies play a vital role in this activity.

Time delays in carrying out this activity will result in the company not catching up with the market and the consumers turn to other product varieties manufactured by competitors. Product planning is an activity that is required for continued success of the business of the company.

Global Outlook in Product Development (collaborative product development)

Engineering design across the world has undergone tremendous changes. The no. and variety of tools for CAD enables establishment of specialized *design centres* that can be spread across different countries.

Design can be made in one country, products manufactured in another country and sold across many geographical locations worldwide. The outsourcing of design and manufacturing reduces the manufacturing cost. Also, new products conceived in one country can be launched across the globe simultaneously thus reducing the "time-to-market".

Collaboration between Product Designer and Process Engineering

- *First step* for organizing the company for concurrent engineering.

It is highly desirable that the product designer and the process engineer (i.e. manufacturing engineers) to *collaborate* on the design of the product/part to ensure that realistic specifications of the material and quality (i.e. dimensions and tolerances) are specified and that the product is produced as designed by the most economical method.

Organizing for Concurrent Engineering

For concurrent engineering to succeed some organizational changes are required:

(1) The commitment of top-level management to implement concurrent engineering.

(2) The *formation of cross-functional teams/groups* consisting of engineers and managers drawn from different departments (Fig. 1.5).

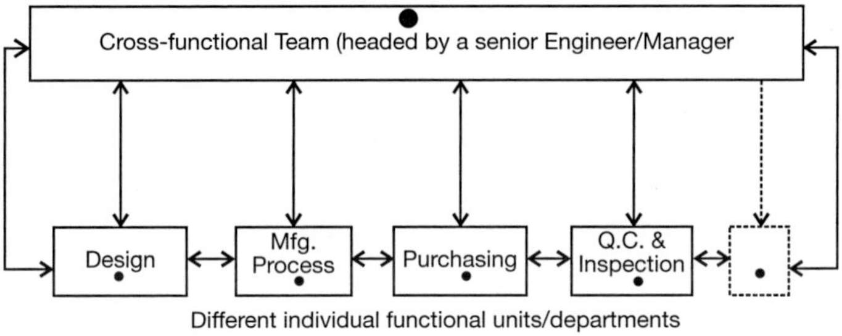

Fig. 1.5: Cross-functional group of people drawn from different units committed to implement CE. The cross functional team consisting of representatives from all functional groups is involved in all the stages of the product life cycle.

The formation of cross-functional team helps in design validation, development of suitable manufacturing process and identification of any new equipment or facility required for the successful implementation of concurrent engineering.

(3) The willingness to share the information amongst employees and teams.

(4) CE will be *implemented differently by different companies* because of differences in leadership, product types and variety, type of customers/market and relationships with vendors and suppliers. This aspect should be considered and required organizational changes are necessary.

Organizational changes required for successful implementation of Concurrent Engineering

Changes required for concurrent engineering	Organizing for Concurrent Engineering Involves
At organizational level	*Multi-functional team or cross-functional team* are to be formed, with emphasis on team objective (instead of department objective).
Frequency and timing of inputs (design changes/ improvements)	Inputs and design improvements required are taken simultaneously from different departments whenever design changes are made.
With regard to *information system*	Willingness to share information amongst employees and teams with respect to design changes made or required.
	Establishing computer systems/ networks (hardware and software) to transfer information to all functional groups quickly and simultaneously whenever new designs/changes are made.
Physical location of functions	To be located in one area, as far as possible.
Central data base is required wherein information can be easily accessed, interpreted, changed and approved.	Computer systems such as CAD/CAM is to be established for implementing concurrent engineering.

Concurrent Engineering: Tool Box

There are many *techniques* (*or tools*) that can be applied/ used for overcoming the difficulties that may arise in developing and implementing concurrent engineering. These are collectively referred to as *Concurrent Engineering tool box*. A particular technique is identified and used depending on the circumstances and requirements. The techniques are in fact, *Problem-solving techniques*.

Tools or Techniques for Implementing Concurrent Engineering

- Computer assisted techniques: CAD (computer aided design), DFM (designing for manufacturing), CAPP (computer aided process planning), and CAMA (computer aided manufacturing and assembly). Proprietary hardwares and softwares are available for this purpose
- SWOT (Strength, Weakness, Opportunities and Threats) analysis
- QFD (Quality Function Deployment)
- Fish-bone diagram for identifying cause and effects
- CPM (Critical Path Method) and PERT (Programme Evaluation and Review Technique)
- Precedence Diagram
- Morphological analysis*
- Mind-mapping (or concept mapping)
- Porter's analysis of manufacturing competitiveness
- AI (Artificial Intelligence) and Expert System (ES)
- Brain storming
- Edward De Bono's Lateral Thinking
- Value Analysis and Value Engineering

These terms are explained in Glossary of terms at the end.

Morphological Analysis*

It is a method of identifying and investigating the set of possible relationships or "configurations" of a given object (or product), concept or problem complex.

It is a 3-D matrix of a problem complex (e.g. consisting of attributes: "Colours", "Shape" and "Size"). By analyzing this, "*Morphological box*" or "*Zwicky box*" possible configurations of product can be generated. A suitable set of combinations can be chosen. This technique is useful in *new product design and creative designs* can be obtained.

Evaluative Questions

1. What is concurrent engineering? How is this concept different from traditional sequential engineering?
2. Give one important definition of concurrent engineering. Underline the keywords in the definition and explain the meaning of the words or term.
3. Is *simultaneous engineering* and *concurrent engineering* one and the same? Say Yes or No. Substantiate your answer.
4. What is meant by "Cross-functional team"? How is it formed? What is its work?
5. Write three benefits of concurrent engineering.
6. What is Manufacturing Lead Time? What is the influence of concurrent engineering on manufacturing lead time?
7. What is the importance of collaboration between product designer and process (manufacturing) engineer?
8. What organizational changes are required for effective implementation of CE?
9. What are the various tools/techniques available for overcoming the difficulties in concurrent engineering?
10. Explain *one* technique indicating how is it applied in concurrent engineering.
11. What is the meaning of "Manufacturability" or "Producibility"?
12. What is meant by "Time-to-Market" and its significance?

2

Use of Information Technology

Computer in Manufacturing Systems

The introduction of computers in industry is an evolutionary process. It begins with the application approach followed by data entry integration, distributed processing and data storage. All the phases of manufacturing work in coordination and in communication with one another. All the manufacturing activities are tied (integrated) together and are using the same information. This information is retrieved whenever needed, synchronized, displayed in the computer terminal and ready for use without any preparatory or additional work.

Computers and Information Technology Systems facilitate automated planning and execution of all related manufacturing activities in a plant.

IT is a change enabler to transform businesses into vibrant enterprises: *Examples of how IT and computer systems have entered the engineering enterprises*:

- Engineering design is supported by technologies and specialized tools such as CAD, CAM, CAE, CAI and CAPP. (resulting in benefits such as better/optimum design, reduced cost and reduced time to market and improved efficiency of manufacturing operations)
- Designs are made available at different manufacturing units, over internet/intranet, instantly avoiding unnecessary delay and concurrent manufacturing by different units.

- Quick design changes/modifications can be made and presented on the computer terminals of vendors, even at remote locations.
- IT enables implementation of concurrent engineering in the industries.
- Information Technology strengthens the concept that "manufacturing/use of Information Technology (IT) and IT support production flow of components is basically information flow across different work centres."

In a manufacturing plant various functions carried out are:

- Product Design
- Manufacturing/Processing (Process Planning)
- Production Planning and Control (PPC)
- Inspection, Testing and Q.C. (Quality Control)
- Business functions (sales & marketing)

Many companies make hundreds of different products; each product consisting of a no. of individual components.

Need for IT Support

The information concerning product design, and data is documented by means of component diagrams, specifications-fits and tolerances, and materials required to produce and assemble them. The task of coordinating all these activities required to make the parts, assemble them and deliver the products to the customers is a *complex job*.

The common denominator that drives and integrates the manufacturing functions is *product data information. Material flow* from one stage to another stage is basically considered as *information flow*. We need a well-coordinated MIS (Manufacturing Information System), for smooth functioning. Also, information flows back and forth between different stages in the spectrum of manufacturing activities.

Manually retrieving every time such information whenever it is required is cumbersome and time consuming. Hence, IT support is required for quick and reliable data base and product data management.

> **IT is an enabler**
>
> Manufacturing enterprises need to get better at:
>
> – Storing, Managing, Reporting, Analysing, and Reusing their Production and Design data.
>
> Data should be leveraged to build AI based platforms that can transform customer satisfaction into delightful experiences.
>
> IT enables to shift from systems of record to systems of engagement.

Information Technology devices can execute simulation design of automotive cars, and perform precise remote measurements. IT enables the promotion of innovation, quality and manufacturing competitiveness.

Solid Modelling

Geometric Modelling: Geometric modelling uses a CAD system to develop a mathematical description of the geometry of an object. Parts are defined as solid objects. The mathematical description is called a "model" and it is stored in the computer memory. This enables the CAD engineer to develop the image of the model on a graphics terminal of the computer to perform certain operations such as:

- Cross sections can be cut through the model showing the internal details,
- Creating new geometric models from basic building blocks available in the system,
- Moving the images around the screen, and
- Zooming in on certain features (part features, e.g. holes, corners) of the image.

These capabilities enables the CAD engineer to construct a model of a component/product. Existing models can also be reviewed.

Primitives such as cubes, spheres, pyramids and cylinders *are combined* to build up the solid model of the part.

There are *two-types of geometrical modelling* in CAD system.

(i) 2-D modelling (2-dimensional): basically used for CAD drawings and drafting.

(ii) 3-D modelling (3-dimensional) for constructing various views of solids (solid modelling).

Solid models are constructed in two basic ways.

1. Constructive representation (C-rep) or building block approach (most industrial parts are made by this method).
2. Boundary representation (B-rep) used for parts with complex contours (Turbine blades).

In *Solid Modelling*, the operations and transformations on the Model are done by the system according to the instructions. *3-D solid modelling* is helpful in the visualization of the objects since they can be displayed in various views and from different angles.

The geometric modelling created can be stored in the CAD system.

Certain mechanical strength properties of the solid object (component) created can be calculated and how they will perform (deflections, if any) in real-life actual loading conditions can be studied.

The models created can be reviewed/evaluated and the required model can be selected. Fig. 2.1 shows a solid model of a ball bearing.

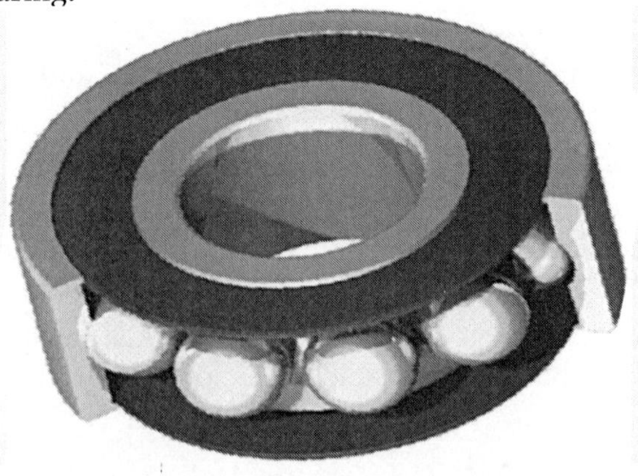

Fig. 2.1: Solid Model of a ball bearing

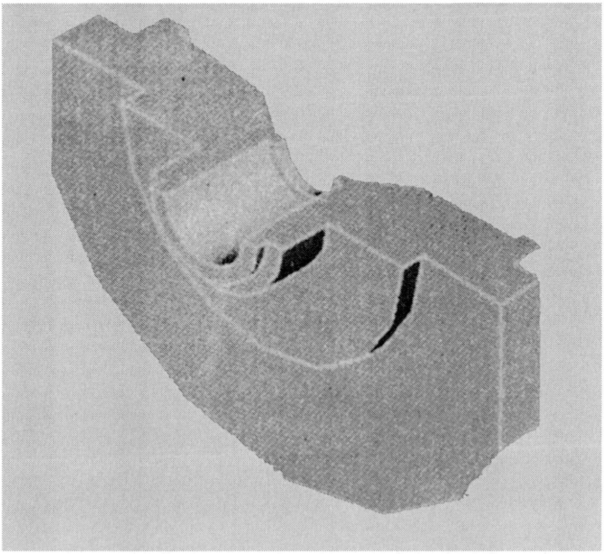

Fig. 2.2a: Solid Modelling of a casing cover showing internal details on a cut-section

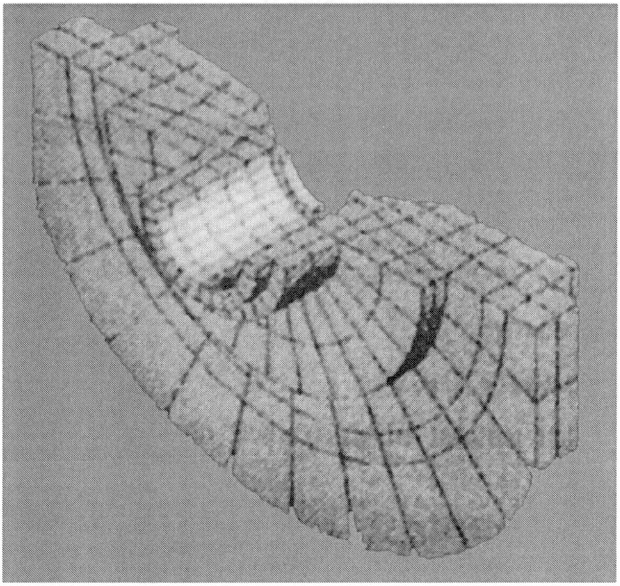

Fig. 2.2b: Finite Element Analysis of the casing cover

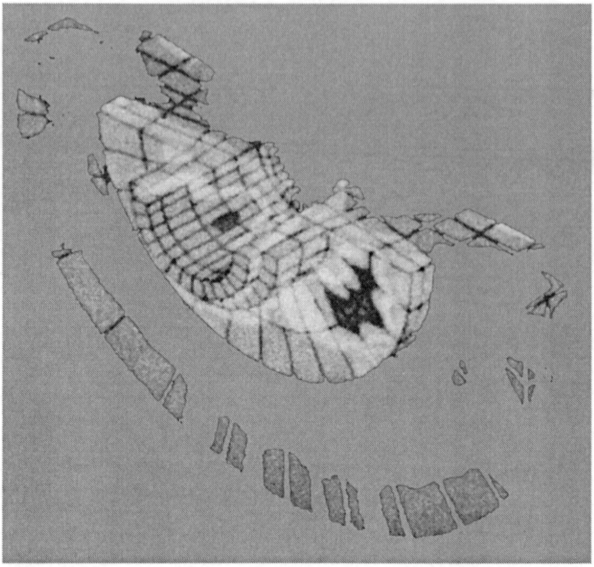

Fig. 2.2c: FEM showing stress contours of casing cover (i.e. how parts designed will respond to load conditions)

Fig. 2.2a, 2.2b, and 2.2c shows certain screen shots on the computer that have been created by CAD system. Component chosen is casing cover for a fibreglass reinforced pump.

On the solid model constructed, engineering analysis such as Finite Element Analysis can be carried out using information such as part material, MI (Moment of Inertia) and other properties that affect how the part responds/behaves under service/load conditions.

Product Data Management (PDM)

Manufacturing engineers need to access a variety of information regarding the products manufactured by the company. PDM enables the engineers to *access these information efficiently and effectively in time*. Relational Data Base Management system is used in PDM.

The Product Data Management is to be updated periodically and stored such that it can be retrieved immediately for study and revision, if any, required. Product Data Management is basically creating a Master File concerning the products/parts made and how can it be used.

The following activities are carried out in product data management:

(1) Documents Concerning the Product Details are Collected, Characterized and Classified

Classification of products are identified based on:

- Assemblies, sub-assemblies, components, their material details, dimensions, tolerances (part geometry) and surface finish requirements. (Bill of materials, parts name/code no. are useful in this regard.)

The data base concerning the individual products are classified and listed on the basis of:

- 2D Drawings/3D models (this constitutes the part geometry)
- FEA (Finite Element Analysis), i.e. how the product/assembly behaves under load/service conditions
- Process planning and tooling requirements
- Test Data
- Technical publications/product brochure
- Spread sheet files
- Circuit diagram (in respect of electrical, electronic and network products i.e. cell phone)
- Feedback from customers, if any. (Fig. 2.3)

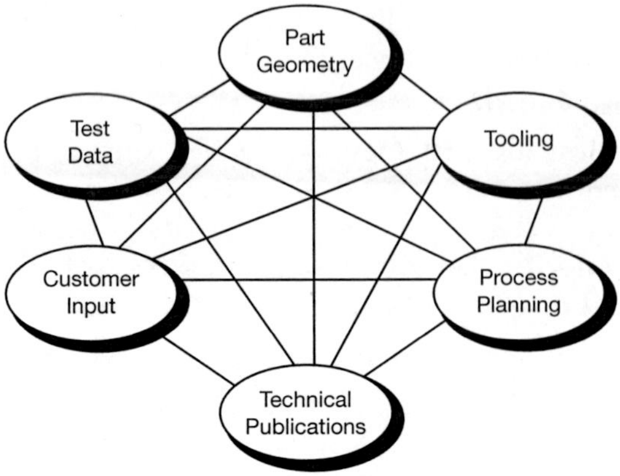

Fig. 2.3: Different aspects that constitute product data. Efficient and timely use of these data enhances the *Product Data Management* and its organization

Definition of PDM

PDM is the integrated and computer supported (IT supported) management of *a variety of information* developed by different departments during every phase of the *product life cycle*, viz.

- Ideation (requirement analysis)
- Design and development
- Production
- Delivery and logistics support

for effective and easy access of the data.

In essence, PDM integrates various activities of product development and product life cycle.

How is PDM used?

Concurrent engineering team members can share these data base/information for their discussion, decision-making, any changes/revisions to be made and actions to be taken.

This enables the task of product development easier and acceptance by all members of the group.

(2) Organization of the Product Data Management System

- A considerable amount of customization is required in PDM since the requirements of the company or organization are different, and it needs to be continuously developed, refined and updated.
  - All data are meticulously *captured in electronic form*, any data changed/revised are recorded with date of revision and who authorized it.
  - Guidelines as to how the data/information is stored, retrieved and shared across different functional groups are also worked out.

(In the early days, "*Master Files*" and *design changes are manually recorded* and "*Project Folders*"/Paper based packets maintained. Now-a-days, advancements and IT support, specially designed softwares enable these data to be stored electronically and retrieved at the press of the button).

- These data can be moved or transferred from one dept. to another in an organized and secured way, maintaining the *data integrity and security.*

Benefits of PDM

(1) The required *data/information* concerning any product or its components (viz. product data) can be *easily and quickly retrieved* and updated by authorized persons for taking necessary action, viz. *any design changes* required.

(2) It takes *less time for engineering design changes* and for preparing tooling drawings and elements.

(3) Redesigning and reworking of parts is reduced to a minimum.

(4) Reduction in "*Time-to-Market*" (it takes less time for new products to reach markets for sales).

(5) Productivity of the design team is increased, since up-to-day information concerning the product can be easily and quickly retrieved, hence designers time is made more productive.

(6) *Creativity of the design team is also increased.* The design team is relieved of the efforts in accessing the routine information. Hence more time is available for design team in designing of new product ranges with creativity and innovation.

(7) Accuracy values, latest version of design and its modifications are made available easily, enabling *better quality control* and less reworking and scrap.

(8) Above all, the product information is shared amongst all stakeholders (in particular, design team, process planners, Tool engineers) facilitating team working in different offices/places of the company *facilitating collaborative manufacturing*.

Softwares for CAD, CAM and CAE, and for PDM and ERP

All CAD, CAM and CAE softwares are capable of supporting PDM in various divisions of the company. However, customization is required to cater to the specific requirements of the enterprise.

PDM (Product Data Management) deals with all aspects of the parts, their geometry, sub-assembles and assemblies at the *product level* whereas ERP (Enterprise Resource Planning) is more broader operating at the *company level* incorporating functions such as Financial and Business Planning, Budgeting, Sales and Marketing. Both have *complementary functions*.

Earlier PDM and ERP softwares were developed as separate entities. Now with advancements in IT and IT support, both the functions are made available in an *integrated manner*.

PDM supports concurrent task management and concurrent engineering as well. Also, PDM is useful when *process audits* are performed for conformance to international quality management standards such as ISO 9000.

Collaborative Product Commerce

Collaborative product commerce is a set of principles, policies and strategies adopted by MNC (Multi National Companies) and large corporates for managing their business operations by collaborating with other companies and industries in certain specific activities.

Collaborative product commerce is emerging as a new trend in automotive, electronics, communication equipment and devices, garments and consumer industries. Design of a new product is conceived and finalized in a design centre in one country or company, design details and other product details are made available on the computer terminals through internet and Web based systems and development and production takes place in another company/country, and components are sourced and manufactured in a third company, assembled and sold across the globe. Industries collaborate with each other sharing their expertise for mutual benefit.

(e.g. Countries with expertise in design capabilities such as USA, Germany, Italy, France and Japan, get the parts outsourced from countries such as India and other South Asian Countries, Malaysia and sold across the world.)

Individual enterprises should maximize their performance in these activities to excel and to become competitive. However, they must concentrate on one or two aspects of their business operations to gain competitive advantage utilizing their competencies (Unique Selling Proposition).

Strategies Adopted by Companies for Collaborative Product Commerce

- Manufacturing to order only (i.e. in contrast to manufacturing to stock/store).
- Setting up reliable and faster ICT (Information and Communication Technology) networks/systems so that design details can be shared by designers and engineers working in different locations, reviewed and approved for manufacturing.
- Companies should develop expertise amongst its tool engineers and set-up engineers/operators (machine setters) for quick change over of set-ups of machine tools to produce one part to another part depending on the orders received and supplied quickly.

- Establishing efficient supply chain networks and logistic arrangements for smooth transport of materials, parts and sub-assemblies from one geographical location to another for distribution to the customers.
- Companies engaged in collaborative product commerce tend to keep their inventory small and they have leaner organizational structure.
- They have to maintain high standards of delivery schedules so that the parts/products reach the distributors quickly.
- Collaborative companies must understand the business trends that govern global economy and commerce and they should view the entire world as one big market place (catering to the customers worldwide).

Advantages of Collaborative Product Commerce

(1) Collaborative commerce leads to new business opportunities.

(2) Locally available material resources and labour are used which helps in keeping the manufacturing cost to a minimum.

(3) Collaborative product commerce enhances product design, product range and variety, quality and reliability and speed of delivery utilizing individual companies and capabilities.

(4) New products and product variety are made available to consumers at an affordable cost, which otherwise may not be possible.

(5) Collaborative commerce also brings in Foreign investment and improves employment opportunities for local population.

(6) Global commerce and business leads to international trade, economic growth of countries and prosperity of their citizens.

Artificial Intelligence (AI)

Artificial Intelligence is a branch of computer science dealing with symbolic, non-algorithmic methods of solving problems.

There are many definitions of AI

- AI is the development of computer systems capable of *initiating human reasoning* and perception (thinking).
- AI is the process of *symbolic information* such as concepts, knowledge and relationships.
- AI is the study of techniques for *solving very difficult problems* in a short-time.
- AI is concerned with *designing intelligent computer systems* and *machine learning.*
- AI is concerned with the *modelling of human intelligence* to do things at which people are better.

AI is used in applications, methods, and techniques of *constructing man-made intelligent machines*, (might be used in the factory of the future, FoF).

A typical use of AI is in *automation systems* driven by manufacturing production data. Data automation systems are used in CAPP (*Computer Aided Process Planning*) and in training. (Such functions otherwise require extensive thinking, are labour intensive and are time consuming). AI in also used in *scheduling* and in *cost estimation.*

AI techniques are used in the development of Expert Systems, knowledge representation and in automated reasoning.

Devices in which AI is used:

- TV's (recording a favourite programme)
- Washing machine (timer, making warning signals, buzzers); switching from one mode/cycle to another automatically)
- Most of the Mechatronic devices
- A/C, Temperature Controls
- Automobiles; giving signals/warning messages, self-locking

- Auto-pilots used in air-crafts; automatic starting of engines and shutting of engines
- Automatic switching on and switching off of lights
- Micro-wave Ovens: timer, temperature controls

(Now-a-days, most of the devices, have built-in intelligence)

Application areas of AI

- Logistics arrangements, sequencing, programming
- Decision-making/reasoning
- Game playing, chess, etc.
- Couper vision, machine vision, machine learning
- Face recognition, voice-recognition, colour recognition
- Robotics
- Simulation, modelling & optimization
- Defect/failure analysis
- Medical diagnosis information search
- Automatic coin sorting m/c
- Translation from one language to another (to some extent)
- Adaptive structures
- Expert systems

Expert System (ES)

Expert system is an application of *Artificial Intelligence*. ES is applied in many industries such as medicine, Air-Traffic Control (ATC), manufacturing CAPP (Computer Aided Process Planning), electronics, information processing, training, petroleum refineries, defect analysis, etc.

ES is basically a computer programme that uses the organized knowledge data base of the human experts. Large volume of information and knowledge (of human experts) is entered into the computer systems coded as a series of rules and applied to a specialized structured base containing information about a real or hypothetical situation or a problem.

By taking action specified by the machine the *computer simulates the behaviour of human experts* in confronting a similar situation/problem.

Expert systems are provided with certain basic rules and they are taught (i.e. to teach them) how to *recognize patterns* and gives a solution to a problem.

Software Development Cycle for Concurrent Engineering involves different phases or stages:

(1) Requirement analysis

(2) Preliminary design of software

(3) Detailed design of software

(4) Programming and coding

(5) Testing and Installation (in departments and in a centralized location)

(6) Training of employees and staff

(7) Maintenance

(8) Correcting errors not previously detected.

It has been found, generally, about 5-7%, of annual expenditure is incurred in the maintenance of softwares and in modifications.

Co-design

In line with the principles of concurrent engineering, while software development takes place, required computer systems/ computer net works required are also identified and selected (Co-design: hardware/software).

Mechatronic devices such as home appliances (e.g. Dish washers, and washing machines) are equipped with the software along with hardware, in the first instance itself when they are designed and developed (Co-design: software/hardware).

Hand held devices (such as calculators, cell phones) are also designed using the concept of hardware/software co-design.

Evaluative Questions

1. IT is a "*change enabler*". Substantiate this statement.
2. "Material flow" in a manufacturing plant is basically "*information flow*" across different work centres. Explain this statement.

3. What is the *need for IT support* for successful implementation of concurrent engineering?
4. Explain how solid modelling is created in a computer system.
5. What is achieved by solid modelling of a part/product? (How is solid modelling is useful in new parts/product development?)
6. Give *one definition* or statement explaining AI (Artificial Intelligence).
7. Give *three applications of AI* in manufacturing engineering.
8. What is *Expert System*? Where is it used?
9. What is your concept of *Co-design* in the context of software development?
10. In *Product Data Management*, what product data is implied?
11. What are the steps involved in *product development* cycle?
12. What are the various details that govern the term "*Product Data*" in Product Data Management?
13. What are the *benefits of PDM*?
14. What is the relationship between *PDM* (Product Data Management) and *ERP* (Enterprise Resource Planning)?
15. Explain how advancements made in IT enables efficient PDM in an enterprise.

3

Design Stage in the Life Cycle of Products

Product Life Cycle

The pattern of product life cycle is shown in Fig. 3.1. This pattern is almost same for all products (domestic and industrial products). The product life cycle is shown in terms of sales volume and profits. The length of product life cycle span is about 15 to 20 years; with a payback period of about 5 to 8 years.

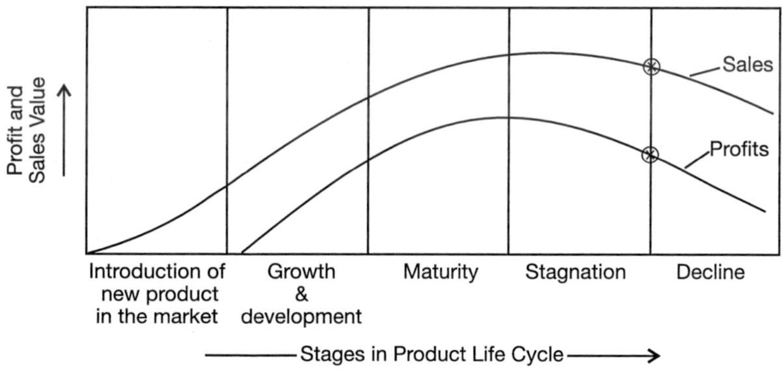

Fig. 3.1: Product Life cycle, showing different stages/phases
⊗ At this stage, *new product ranges* with additional features have to be developed and *introduced* in the market, to remain in the business

Product Life cycle has five distinct phases:

(1) *Introduction* of product in the market
(2) *Growth and development*
(3) *Maturity*
(4) *Stagnation* (or saturation)
(5) *Decline*

(1) *Introduction* of any new product is based on the felt needs of the consumers and market analysis & research, and the product is *commercially launched* in the market. Feedback is continuously received and reviewed and any design changes required are incorporated. Introductory phase is complete when it is known that the product is working satisfactorily.

(2) *Growth and Development* is ensured through advertisement, and sales promotion campaign. Sales and profits begin to rise in this phase.

(3) *Maturity*: Sales volume and profits further rises. Production in the factory also gets stabilized. A fairly constant market share is maintained through sales and advertising.

(Market share, expressed in %, indicates the share of the company's sales volume with respect to the total sales volume of all companies put together).

Competitor's activities are studied and sales volume is maintained.

(4) *Stagnation*: At this stage, market gets saturated, sales reaches their peak value and levels off. There is no additional demand. Sales and profits start decreasing. This may be due to competition in the market.

(5) *Decline*: Sales volume and profits starts falling down, may be due to introduction of new products by the competitors; also foreign companies penetrating domestic market. Reduction in profits may be due to rise in cost of raw materials, wage increase and inflation. Under these circumstances company may begin to stop the production. If the company wants to remain in the business, *new product ranges with new and additional features*

have to be developed. For this to happen, the activities in the company must be synchronized with the development of new products. (Here, product life cycle is considered from business/ company's overall perspective.)

The time span covering all these five phases is about 15-20 years. Management must be aware of these aspects.

Product Life Cycle, in fact, is made of many individual processes. The cycle as a whole is divided into *two broad phases*.

(i) Premarket phase, and

(ii) Market phase.

Individual activities in these phases are:

Premarket Phase	Market Phase
– Idea generation – Idea evaluation – Feasibility analysis – Technical R&D – Product/market research – Preliminary production (Trial batch) – Market testing (and feedback from users, if any) – Commercial production	– Product introduction in the market – Market development – Rapid growth – Competitive market (competitors selling a similar product) – Maturity – Declining of sales – Product being taken off the market – Introduction of new product with more/ additional features

Innovation leading to the development of a new product in the competitive business/market is a complex, costly and time-consuming process. Manufacturing engineers and product designers play an important role in this process.

Uses of Product Life Cycle

Once the pattern of product life cycle is understood, it becomes a valuable management tool. Uses of product cycle are:

- It can provide a framework or basis for developing new products. This will ensure that the new product is ready to launch when the existing product reaches a "declining" stage.
- It can ensure that sales and marketing efforts are intensified at the appropriate time.
- It aids in financial forecasting and necessary funds are budgeted as different phases depending on the requirement.
- It helps in analysing the performance of existing products in the market and the need for developing and launching new products.

Modality of Concurrent Engineering Design

This is the way by which "Product Design" and "Process Design" are considered together concurrently and simultaneously in the initial stages itself during product design and product development.

Product Design refers to the design of the product to meet certain functional requirements and customer needs. This job is primarily done by design engineers. Design is fundamental to the effectiveness of every product made.

Process Design refers to the manufacturing processes through which the parts are taken through following a definite sequence of operations from raw material stage to final stage. This job is primarily done by process engineers or manufacturing engineers.

Product Design

- A product is designed to meet certain functional requirements and to satisfy customer needs. At the same time the product should be
  - of high quality and reliability (durability),
  - safe to handle and use,
  - aesthetically appealing, and
  - easy to service and maintain (easy access to oiling and greasing).

- During product design stage new materials, new product trends and new technology available (such as use of CAD packages) has to be explored and maximum benefit derived from these developments to realize low cost and high quality.
- A product consists of
  - assemblies and sub-assemblies
  - bought-out components (standard components) procured from market.
- If the product/equipment is to be made suiting the customer's special requirements, broad specifications are given by the user. Discussion should be held by product designers with the customers to understand the special requirements and what are his expectations.
- The cycle of events (product life cycle) begins with the sales/marketing department based on the market studies. *Product ideation* occurs, designs are conceived and one or two pieces are made (prototypes or trial pieces) for testing and to study their performance.
- The product design is documented by means of component drawings, specifications and materials required to make the final product. Any special requirement with regard to heat treatment or finish are also specified.
- Traditionally manufacturing dept./engineers are not consulted at this stage. But, concurrent engineering demands that product designers consult and discuss with the process engineers/manufacturing engineers regarding the details of product designed to understand the special requirements of parts or assemblies.
- Design changes, if any required, are identified without sacrificing the functional requirements. Such changes are studied and agreed upon by the product designer.
- Cost estimates are prepared to know the anticipated price of the product.
- After the completion of the design a presentation is made so that top management understands the product details

and requirements. Budget provision is made for purchase of any new equipment required. Authorization is given to start the production.

What is Process Design?

- Next stage is process design. Process design primarily means identifying the processing sequence required to carry-out individual operations and assembling of the components and sub-assemblies. This sequencing is as per the technological sequencing required.
- Product details are studied by the process engineers/ manufacturing engineers and a Route sheet (or known as process planning) is prepared.
- This process planning lists the sequence of operations, on which machine tool or equipment, it has to be carried out, the tooling required (Jigs, fixtures or cutting tools) and the machining conditions are specified.
- Machining conditions are chosen so that the tolerances specified on the production drawings are achieved.
- What lot size is to be produced is decided as per the production planning and schedule worked out.
- Ensure that all steps are taken to keep the manufacturing cost to a minimum.

Factors affecting Process Design

1. Total quantity of production and lot size.
2. Production rate per hour or per week.
3. Target cost of the product as envisaged by the costing dept. and sales/marketing dept.
4. Availability of production equipment and their capabilities to achieve prescribed tolerances.
5. Technology available (special purpose machines, SPM, Robots, CNC Machines or general purpose machine GPM only).

This is how product design and process design is integrated and modality worked out by the design team and manufacturing engineers for concurrent engineering design.

Product Engineering

Produce engineering is mainly concerned with product design and how is the product to be manufactured? Product engineering will consider how is the product to be manufactured at minimum cost.

> *Product Design* is documented by means of component drawings, specifications and list of materials, and parts required (bill of material).

Step involved in product engineering are:

- *Design of components*, mainly involving sketching and drafting work (*conceptual design*).
- *Preparation of specifications*, mainly critical dimensions, limits, fits and tolerances, and finishes required, also materials are specified (*detailed design with engineering drawings*).
- *Maintaining product standards and quality* to ensure product performance at minimum cost, also standardization of parts and components.
- *Product Testing* for performance in actual user condition/ or in service condition, either in the company laboratory or in the field. Ensure whether product performance is guaranteed and there are no defects. Qualified test engineers will carry-out this task.
- *Servicing and maintenance*, whether serviceability requirements have been taken into consideration; i.e. use of "modules" or standard parts for quick and easier maintenance.

Functional Groups in manufacturing and their corresponding activities are shown in Fig. 3.2.

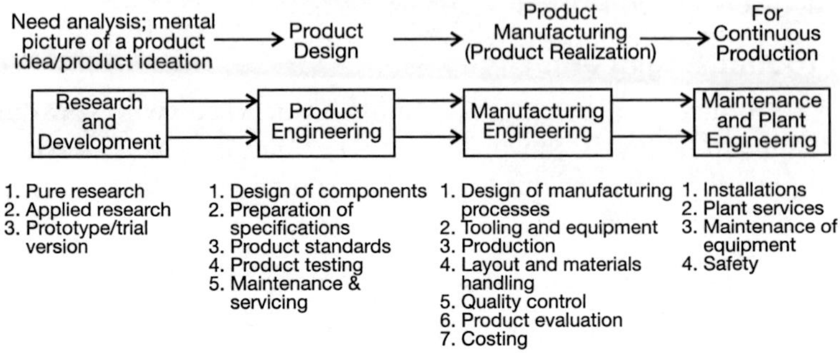

Fig. 3.2: Grouping of the engineering functions in manufacturing and the corresponding activities

Concurrent Engineering in Optimal Design

The balance between function, quality, reliability, appearance and manufacturability (i.e. production) should be taken into consideration in the *optimum design of a product*. Optimum design should also be such that it results in *affordable* cost to the user.

Optimum design requires a knowledge of materials, production processes and the needs of the user. If it is not possible to achieve optimum design in the initial stages itself, the following strategies may be adopted.

– Redesigning an existing product,

– Make design changes,

– New product design itself, if the above are not possible.

New product design is usually based on the identification of a need that the existing design not fully able to meet or able to meet only partially.

The "*Design-to-cost*" concept emphasizes the importance of design on the ultimate cost of the product.

Designing for reduced cost is an important aspect when changes in design of existing product or in designing a new product are undertaken.

When considering cost, the designer has to examine *five main aspects* of the product:

1. Materials to be used
2. The production processes to be employed
3. Quality required
4. Maintenance cost
5. Operating cost associated in using the product (i.e. energy cost)

The designer must be fully aware of the implications of his design on the final manufacturing cost. Optimal design must lead to economical cost to the user. In this aspect, concurrent engineering is a very useful tool/technique.

Opportunities for Manufacturing Enterprises

Manufacturing is a production business that is intended to make a profit. Manufacturing is an important economic activity for any nation for creation of wealth. Manufacturing sector plays an important role in the economic and social development and in job creation of any country and in international trade. The global competition among the nations has resulted in the development of new materials, processes, equipment and technology.

This means there are enormous opportunities for manufacturing enterprises. A wide variety of products and goods can be produced by industries. These are:

- *Consumer goods*: Furniture, domestic appliances, soaps, dresses, sports equipment, textiles, cell phones.
- *Industrial goods*: Machine tools, power operated tools, hand tools, materials handling equipment, gauges, measuring instruments, construction equipment.
- *Processing equipment*: Food processing equipment, chemical plants, pharmaceutical machinery, refineries, paints, manufacturing enterprises have enormous potential in job creation and employment and in export marketing.
- *Network products*: Mobile/cell phones, electronics components production and assembly and telecommunication equipment.

Whether to choose diversified product or product range or specialized production?

This varies from one industry to another. The decision:

(1) Depends on product or product range in which the company is already engaged in.

(2) Whether the industry wants to concentrate in specialized products to gain competitive advantage? (USP: Unique Selling Proposition?)

(3) Availability of skilled labour required in making specialized products.

(4) Availability of components material/parts (whether to be imported or locally made available?)

(5) Whether the production is exclusively for domestic market or consumption or for export trade?

A judicious decision should be taken by the industry and a balanced approach is desirable. Also, the company has to study the market demands on a continuous basis and remain *agile* to make necessary changes in products range or to develop new products to continue to be in the competitive business.

Real-Time Constraints (in implementing Concurrent Engineering)

"*Real-Time Constraints*" mean the difficulties (obstacles) that occur while actually implementing CE from time-to-time.

Constraint or Difficulty	How to overcome them?
• Management has not fully understood the concept of concurrent engineering.	– The cross-functional group manager/leader must explain to management about CE and its benefits.
• Management is not fully committed to CE as a new approach to manufacturing competitiveness.	– -do-
• New equipment/new facility required for CE is not put in place for new process planning/ computer systems required.	– Discuss with management and convince them about the need for this new equipment/facility required for process planning and for implementing CE.

Constraint or Difficulty	How to overcome them?
• Design engineers are very rigid w.r.t. materials specified (which are difficult to procure) and the tolerances specified and are not willing to relax them in view of safety requirements of the product.	– The cross-functional group leader/manager must discuss with their engineers about the equivalent material in the market which will still meet their requirements.
• Vendors are not able to supply the parts in time for assembly.	– Workout a suitable arrangement with the vendors. Engineers of OEM plant should visit suppliers site and help their staff to streamline their production methods and guide them in Q.C. and quality improvement. – Encourage them to adopt the practice of "Kitting"/Pre-assembly of components/part before despatching the items. – If possible arrange a training programme for their engineers/staff regarding concurrent engineering being practiced by OEM plant.
• Process planning records (route sheets) are not updated incorporating the design changes whenever they are made. • Operating staff do not understand the new approach, (CE), being introduced in their plant.	– Engineers/managers of cross-functional group should conduct training programmes to their operating staff about the various aspects of concurrent engineering.

Evaluative Questions

1. Discuss the necessity of close cooperation between the *product engineer* and the *process engineer* (i.e. manufacturing engineer).
2. What is meant by "*Line Balancing*" and state its importance?
3. What is "*Economic Lot size*"? What is its significance?

4. What is "*Life Cycle Design*" in the context of manufacturing industries?
5. What are the "*Constraints*" likely to be faced in the implementation of concurrent engineering?
6. How concurrent engineering will help in the *optimum design* of components in product design?
7. What are the *opportunities* available for companies/ industries to develop new products in the local market and in foreign countries (global/export market)?
8. Distinguish between "*Product Design*" and "Process Design". Which one comes first?
9. How can a company take a decision as to whether go for diversified product range or specialized production?
10. *List some of the activities* carried out during pre-market phase and market phase.
11. Considering product life cycle, at *what stage new products* are designed, developed and introduced in the market?
12. List and describe the 5 distinct phases in the product life cycle *from the company's business point of view.*
13. What is the *use* of understanding *product life cycle* as a valuable management tool for undertaking *business activities of the company*?

4

Manufacturing Concepts and Analysis

This chapter involves an understanding of the concepts and analysis of various activities involved at all stages in the manufacturing cycle. (Fig. 4.1)

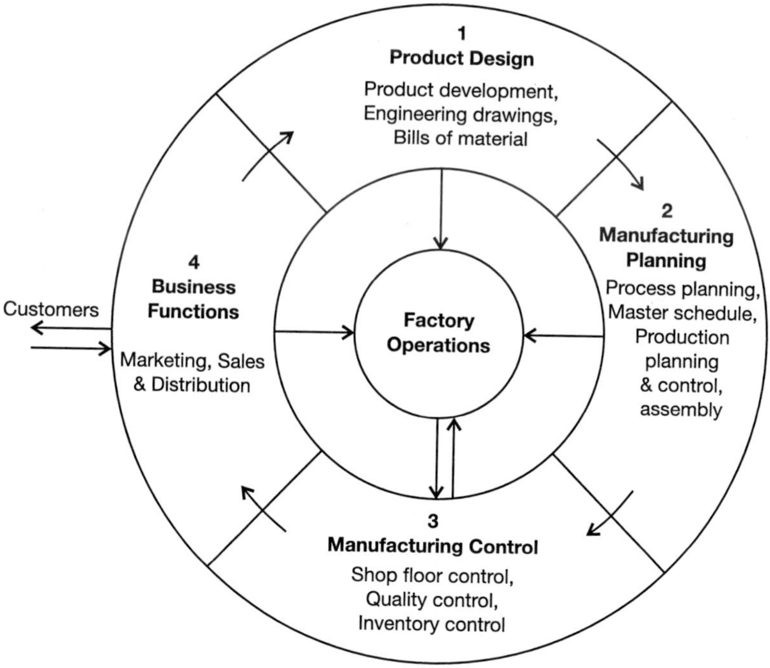

Fig. 4.1: General activities involved in manufacturing firms. (organizational functions within a manufacturing system/firm)

- *Design Analysis*: Leading to effective and robust design for meeting the functional requirements of the product: strength analysis, dimensional analysis, limits, fits and tolerances specified and tolerance build up when parts are assembled.
- *Manufacturability* considering all aspects of product design, materials selected and manufacturing processes involved in an integrated manner in the initial stages itself.
- *Studying the process capabilities* of production equipment/ machine tools available to achieve the tolerances and surface finish prescribed.
- Whether *modern technologies* and recent technological developments have been considered?
- *Analysis of manufacturing processes* w.r.t. various activities/operations involved and their technological sequencing and process planning.
- *Analysis of inspection procedures* to ensure quality control.
- Analysis as to whether as many *standard parts* as possible have been specified so that manufacturing lead time and cost can be minimized (*make or buy*).
- Check whether *standard assembly methods* (Design for Assembly hints) have been prescribed for proper location of parts, fool-proof placing of parts in the assembly, clamping and tightening and desired fits are obtained, so that there is no interference.
- Extent to which *concept of "Modules"* (*kitting*) can be applied during assembly (capability of vendors to supply parts in time); i.e. modular design can be considered wherever possible.
- Analysis as to whether *principles of automation* have been considered in all successive stages of manufacturing cycle.

Manufacturing Competitiveness

Manufacturing has an important place in the economies of all developing and developed countries—both in terms of employment and its contribution to wealth-generation. The rapid growth in the world wide application of advanced manufacturing technology is providing new opportunities for new products, production processes and manufacturing enterprises. This is also increasing the competitive pressures among the industries.

Hence there is a need for robust competitive strategies to be developed by the industries to stay in business.

Manufacturing Competitiveness (Porter's Model)

Five forces driving manufacturing competitiveness as conceived by Porter, M.E. (1980) is shown in Fig. 4.2.

Fig. 4.2: Process driving Competitiveness in manufacturing industries (Porter's Model)

1. Threat of new entrants
2. Bargaining power of buyers (consumers)
3. Pressure from new products (based on Market Survey, R&D, and new product designs)
4. Power of suppliers (sole suppliers/new suppliers) and their bargaining capacity
5. Rivalry among existing firms.

In addition, other factors involved in manufacturing competitiveness are:

- Whether the company has evolved a manufacturing plan for short range (3-5 years) and long range (15-20 years) period?
- Variety, type and product range developed by the company meeting the customers requirements.
- At what prices the products are sold in the market? (Affordability)
- Whether safety, reliability and product quality have been ensured?
- Ability to provide after sales service to the consumers in terms of accessibility and availability of spares.
- Whether company is innovative and continuously introducing new products and services?

Manufacturing System

It is an integrated combination of processes, production facilities, people, organizational structures, information flows, control systems and computers whose *purpose/objective* is to achieve economic production and competitive performance.

The competitive pressure faced by the Indian manufacturing companies are:

(1) Free access to Indian market by exporters from other Asian countries (such as Chinese, Korean, and Malaysian companies).

(2) Rapid changing technologies/services, especially in the area of computers and electronics.

(3) Rising expectations of Indian consumers concerning product variety, product quality, price and servicing.

(4) Competition from local manufactures also.

In conceiving a new product and in manufacturing it, there are *many activities* that take place. These activities are shown in Fig. 4.3 (both engineering activities and manufacturing activities).

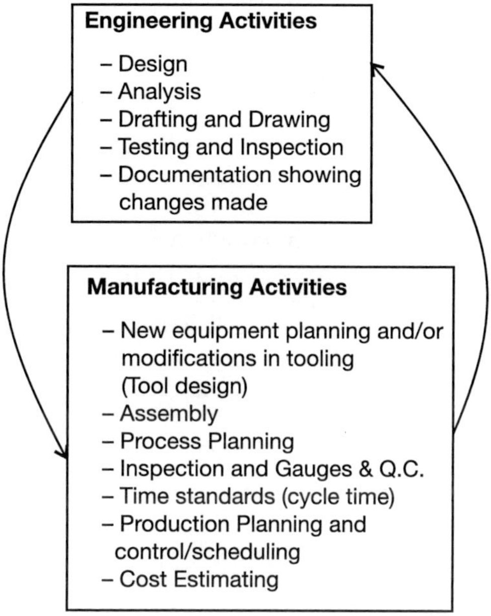

Fig. 4.3: Activities effected by a new design or by the introduction of a new part

All these activities have *cost implication*. Hence, *design changes or design modifications* required should be kept to a minimum and should be carried out *in the initial stages itself*.

Export Competitiveness

Export growth depends on the competitiveness or a country's potential to export to other countries/world market.

Generally, it is stated that India's share of exports is only about 0.2% of total business volume of trade of all countries put together. There is a considerable potential for exports which will bring more foreign exchange entering the country.

Competitiveness in the world market is influenced by factors such as identification of new and innovative products, their design aspects, quality, price (cost) and logistics of transport from one country to another, sales and marketing networks and distribution.

Product design and innovation are the two strategies to win in an international market. Concurrent Engineering plays an important role in this aspect.

> Major model changes as in G.M. (General Motors), USA took five years. With concurrent engineering this was reduced to three years—similar to that of Japanese.
>
> –Carl M. Zienke and Mary S. Spann,
> *Industrial Engg.*, February 1991, p. 45.

Conceptual Design Mechanism

A typical design project (i.e. development of a new product) consists of *three distinct time phases/stages* as per Asimow's Morphology of Design (i.e. Morphological analysis). These are:

(i) Conceptual design,

(ii) Preliminary design, and

(iii) Detailed design.

All the three phases are equally important. You cannot do detailed design without going through conceptual design and preliminary design.

Activities carried out in each phase:

Conceptual Design	Preliminary Design	Detailed Design
– Need analysis of the new product (validation of the need) – Ideation of new product occurs – Carry out the feasibility study – Consider whether design variants (alternatives are possible)	– Structured development of the concept (i.e. new product) – Whether the product meets functional performance and reliability requirements – Product designer gets an idea of assembly, sub-assembly and component parts (standard parts) – *Semi-realization of product idea occurs* – Strength characteristics of materials, type of material required are identified – Typical example of the idea incorporating the functional requirements is worked out.	– Size, capacity and power requirements are worked out – Product features and their arrangements are identified – Major dimensions, limits, fits and tolerances of assembled parts are worked out – Full realization of the product idea occurs – Any specific surface characteristics are required – Aesthetic details are worked out in consultation with industrial designers – Consider manufacturability of the product and major manufacturing technology (processing method required) – Cost estimation (target costing) is done

Excellence in detailed design decides the success and cost of the product.

• Cost of design change at *conceptual stage*, is say ×.	• Cost of design change/ modification at *preliminary design* becomes 10×.	• Cost of any design change at *detailed design* stage increases to 100×.

Hence, any design change must be carefully reviewed at the conceptual design itself to avoid incurring of excessive cost during actual manufacturing/commercial production. Detailed design plans are then considered by manufacturing engineers (process design) in consultation with product design team for batch/mass production.

Checking the Design Process

Concurrent Engineering Design Analysis (or) Hints for effective product design.

Automated Analysis of Parts and Product Design for Concurrent Engineering (Design for Concurrent Engineering).

A detailed analysis of product design/part design and design practices involve the following:

- *Manufacturability or Producibility* of the part or product. This is shown in Fig. 4.4. Manufacturability or producibility of a part involves consideration of the part design/product design, the manufacturing process and the material in a holistic and integrated manner in the initial stages itself, concurrently and simultaneously. This helps in optimizing all the resources to keep the redesign and reworking to a minimum.

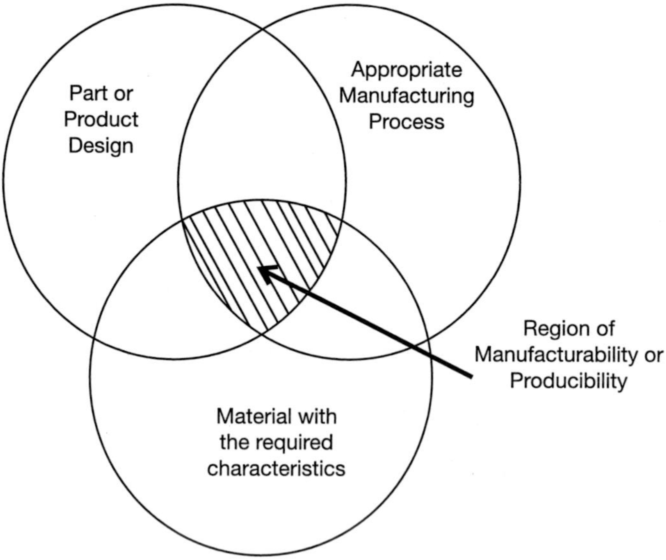

Fig. 4.4: Concurrent Engineering takes into consideration the part/product design, manufacturing process and material *simultaneously* and in an *integrated* and *holistic manner* in the *initial stages* itself.

- *Good design* must be functional meeting the performance requirements, low cost, quality, durability and prossess aesthetic appeal (especially if it is a consumer product).
- As far as possible the assembly or product should consist of *less no. of components*, with modular concept resulting in ease of maintenance, servicing, replacement and repair.
- *Standard parts* and fasteners easily available in the market should be used facilitating standardization.
- *Datum surfaces* of parts should be correctly identified and shown in the shop floor drawings to facilitate machining and assembly. It is preferable that dimensions and their tolerances are specified w.r.t. one common datum surface.
- Critical dimensions should be as few as possible. *Accuracy specified, tolerances and fits assigned* and *surface finish* requirements should be critically analyzed

to reduce manufacturing cost, facilitate easy and quick assembly.

- Location of parts, fixturing, clamping and insertion of parts in assembly should be examined to facilitate easy and quick assembly.
- *Hole sizes*, patterns, and other part features should be standardized to minimize the cost of tool inventories.

It may be noted that the above *product or part analysis* is done in the initial stages itself, involving all the departments.

All the above aspects of *product design* will ultimately lead to *less time to market*, economics of production (less cost), enhancing *manufacturing competitiveness*.

> *Automation* ends the era of low wage.
> India, will need to move up the value chain to remain *competitive*.

Good design practices have been developed for use by product designers. The above guidelines are considered under *Design for Machining*.

Manufacturing Strategy

There are four *principal needs* which any manufacturing strategy must take into consideration, if the company is to be successful and profitable:

- Products must be competitive,
- Quality must be high,
- Delivery of products, servicing, repair, if any, must be promptly attended to, and
- Cost should be minimum.

Any intelligent design of manufacturing system must consider the above four needs, if the company has to succeed and be competitive.

Product must be competitive: A minimum no. of parts should be used in the assembly of products and also rationalization of tooling elements (i.e. use of as many standard parts and tools as possible). Also materials specified should be such that with a minimum range of materials product requirements are met. This will result in less cost and product marketed will be competitive.

Establish suitable manufacturing facility which will meet the current business needs and future activities (for next 3-5 years).

Better interaction with vendors for implementing JIT (Just-in-Time) system. Ensure that vendors supply the required parts at the right time and, if possible, complete sets of parts as packages ("kits")/"modules" i.e. pre-assembled kits so that they can be fitted in the assembly line as soon as it is received. This also keeps the inventory of bought out parts to a minimum leading to JIT approach. For this to happen, engineers of the OEM (original equipment manufactures) must inspect the parts in the vendor site itself before it is despatched to the OEM.

Continuous Improvement in Quality is also required.

Incorporate AMT's and modern machining processes, equipment, computers and advanced manufacturing technologies (AMT) in line with technological changes and advances made.

Reduction in Overheads, inventories (WIP) and indirect labour, and less floor space are some of the valuable gains of intelligent manufacturing system.

Invest in People: Make employees an integrated part of the manufacturing system. At every stage of planning, implementing of new systems and operations, involve the people to the maximum extent possible and take care of their training needs, retraining and special training when facilities are upgraded and modernized.

The above are shown in Fig. 4.5.

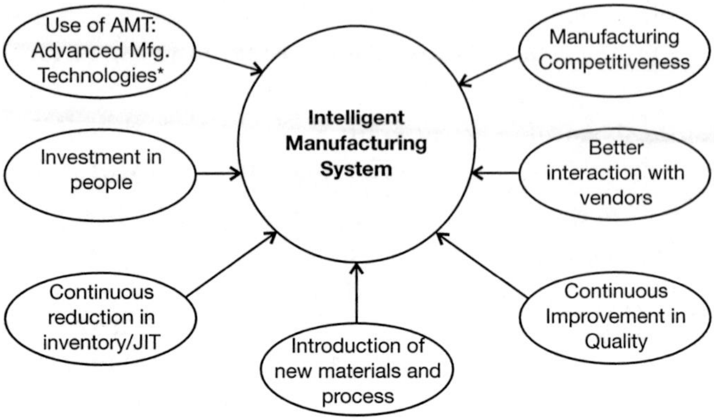

Fig. 4.5: Design of Intelligent Manufacturing Systems Integrates many facets of the enterprise
* AMT means CAD, DFM, CAE, CAM, CAPP and CIM

Need for an Intelligent Manufacturing System: Generally manufacturing systems have often been designed piecemeal (i.e. incremental or gradual development at different times and in different ways) rather than a carefully planned system at the beginning itself in an holistic way.

Hence intelligent design of manufacturing system calls for holistic development at the beginning itself, considering all aspects in an integrated manner. Any intelligent manufacturing system must take into consideration.

- the present and future business needs for competitiveness,
- the quality, reliability and safety requirements,
- responsiveness (i.e. agility) in addressing the needs of the consumers, quickly for any design changes, and
- the cost.

Design for manufacturing (DFM) must result in efficiency and economics of production, i.e. less cycle time, less cost and less rework. Any change required must be accompanied with minimum disruption resulting in maximum benefit.

Evaluative Questions

1. What is meant by manufacturing competitiveness?
2. Explain "*Porter's Model*" of Competitiveness. Is it applicable for Indian industries?
3. What do you understand by the term "*Manufacturability*" or "*Producibility*"?
4. What do you mean by "*Intelligent Design*" of manufacturing system?
5. What is meant by DFM (*Design for Manufacturing*)?
6. What are the different *engineering activities* and *manufacturing activities* to be considered in the *redesign of components/products*?
7. What is the importance of better interaction with *vendors* and *suppliers*?
8. What *competitive pressures* are experienced by Indian manufacturers to remain in business?
9. How is an *analysis of design* of a new product carried out?
10. What in the importance of *less no. of parts* used in a product? What is the significance of *standardization* in this context?

5

JIT System

JIT (Just-in-Time) System

JIT: It is a means of reducing inventory of raw materials and purchased parts. With JIT, a large company that makes an assembled product will require its suppliers to deliver the components needed for the product within a short interval before the assembly process. The interval may be one day or one week. Suppliers/vendors are required to deliver immediately before they are needed in the assembly (i.e. "*just-in-time*"). The benefit to the large company (OEM, Original Equipment Manufacturer) is that inventories are reduced. Instead of keeping several days' supply of parts, the inventories buffer is reduced to one or two days only.

For this purpose, the suppliers should be reliable to make supplies on time, and prior arrangements are to be worked out with them; and coordination with the vendors is also necessary.

JIT and inventory reduction are inter-related. S lots support JIT system. To realize smaller lots, setup time should be as minimum as possible. (This means setup engineers and operators should be well trained in machine tool operations and in tooling.)

JIT's general goal: JIT's general goal is to schedule each material (part or sub-assembly) to arrive just-in-time to be used in the assembly, and this would apply whether the material is produced within the factory or purchased from the outside (suppliers or vendors).

Specific Goals of Just-in-Time (or what it intends to achieve):

1. Produce the product the customer wants
2. Produce the products only at the rate the customer wants them
3. Produce with perfect quality
4. Produce instantly (manufacturing lead time is zero)
5. Produce with no wastage of material, equipment time and labour
6. Produce by methods that leads to the development of staff, engineers and managers.

Toyota (Japanese company) (1980), was the first industry to use this concept in their vehicle manufacturing plant. JIT is also known as pull-system (Kanban system), i.e. preceding work starts only on a specific demand was placed on them. Kanban requires very close cooperation between vendors and main plant.

The JIT concept basically aims at *reducing lot size equal to one* and ultimately to *zero inventory*. However, in practice it is difficult to achieve.

- Japanese companies report inventory turnover ratios in the range of 40 to 60, whereas
- US firms report a ratio of 5 to 8.

JIT system of production was pioneered and perfected by Toyota Motor Co., Japan, (1980) and later followed in USA and other European firms.

JIT is not only an inventory reduction system. JIT production is also related to quality control. The ultimate objective is to operate without any safety stock (i.e. inventory). The advantages or JIT include:

- Continuous reduction in lot size (*The ideal lot size is one*. But it is not possible, in practice),
- Reducing set up times (so that quick changeover from one machine setup to another setup), and
- Cutting purchase orders costs.

JIT may be considered as a part of broader company wide effort which includes:

- Total quality control/total quality management (TQM),
- TEI (Total Employees Involvement), and
- Vendor development.

High levels of quality and reliability are required to make JIT work.

Improvements brought about by JIT in the long run are: – Better quality and reliability, – Cost reduction, – Better space utilization, and – Competitive advantage.

What is Inventory?

Inventory means range of stock of items stored for use and for further processing. These are categorized as follows:

- Raw materials inventory (including bought out components kept in the stores).
- WIP (work-in-process) or semi-finished goods.
- Finished goods inventory.

In addition, there are other types of inventories:

- Tooling systems.
- Spares required for maintenance and servicing of production machineries and equipment (e.g. bearings).

Importance Maintaining Low Inventory

Every effort should be made to keep the inventory as minimum as possible since they represent dead capital and increases cost of production.

JIT system is an approach to keep the inventories down to a minimum (or optimal level).

Work-in-Process (WIP)

It is the quantity of items (components) currently lying/located in the factory that is either being processed or is between processing operations. *WIP is inventory* that is in the state of being transformed from raw material to finished product (units in process). WIP represents an investment by the firm, but it cannot be turned into profit until the processing is completed.

Inventory Control: It is the use of computer data bases to track inventory levels, existing stock and safety levels; and ordering for new stock.

Modular Design

In automated assembly lines, pre-assembled modules (or completed "*kits*") supplied by vendors facilitate quicker and easy assembly.

Assembly of individual parts/components is avoided. Each module may have a maximum of 10-12 individual parts. Assembling these individual components on the assembly line will result in lower production rate.

The Modular concept in design and production is also very useful in trouble shooting, repair and servicing. One has to identify the fault in a particular module and replace it instead of identifying the fault at the individual component level. This concept was first adopted in fabrication and assembly of electronic products and later extended to mechanical products.

The modular concept also *facilitates less inventory and JIT manufacturing*. Arrangements have to be worked out with vendors/suppliers so that the required modules fabricated are inspected at vendor's site itself by OEM (original equipment manufacturers) and their engineers and despatched for immediate assembly directly on the production line. Modules are generally supplied on a weekly basis. This practice is common in automobile/vehicle production now-a-days.

6

Design of Automated Manufacturing System

Automation

The word "*Automation*" is a contraction of the words "*automatic operations*"; it implies continuous and cyclical arrangement for manufacturing, processing, testing and assembly of products as automatically as possible and economically as feasible.

Mechanization

The term "Mechanization" means work is done with the help of machines (machinery, equipment) (i.e. use of roller conveyors (Fig. 6.1) and belt system and materials handling devices and equipment such as cranes and hoists and use of press buttons, switches and setting the machine limits).

Fig. 6.1: Roller Conveyor moving packed boxes from one workstation to another for delivery. (An example of mechanization)

Need for Automation

Doing manually the operations are time consuming and will not fulfill the modern-day requirements of economics of production, low cost, quality and market demands, speed and timely delivery of products and services to the customers. Hence automation is required.

The evolutionary phases of automation management is shown in Fig. 6.2.

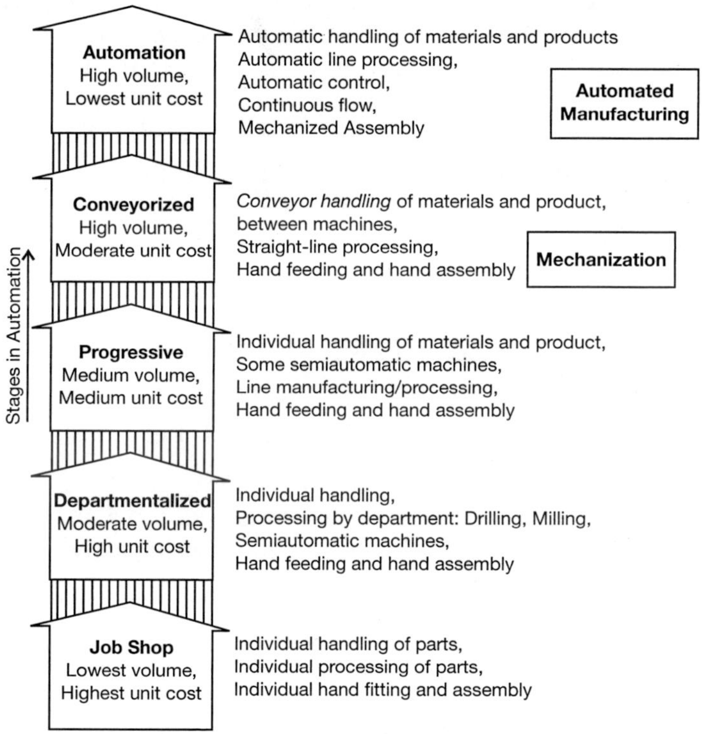

Fig. 6.2: The typical evolutionary phases of manufacturing leading to the automation stage
(Courtesy: Manufacturing Automation Management, Roger W. Bolz, Chapman & Hall, USA)

Development of control devices and large-scale integrated circuits led to the development of automatic controls.

> *Automation* is a technology concerned with the application of mechanical, electrical, electronic and *computer-based systems to operate and control production.*

Some of these individual functions such as processing of raw materials, production of parts by machining can be optimized and automated fully, creating "islands of automation" (sometimes called "*partial automation*" in contrast to "*full automation*"). Full automation is a complex task, involving high initial investment and cost; in some cases full automation, is not feasible and not really required, such as in testing, especially critical dimensions and part features and some assembly operations.

> *Automation technology includes:*
> - Automated machine tools (CNC) machines
> - Automated assemblies
> - Automatic inspection systems (CAI)
> - Industrial robots (Robotics Technology)
> - Automatic materials handling, retrieval and storage systems
> - Feedback control and computer process control
> - Computer systems for data collection, planning, and decision-making to support manufacturing activities and IT support systems.

Distinct stages in the development of an automation system is shown in Fig. 6.3.

Importance of analysis of Product Design

Product design must be robust before start of any manufacturing activity. *Product design* takes into consideration the functional requirements of the product and designing the part/product which meets such functional requirements. Product design is the foundation of all future manufacturing operations. Hence product design is critical and must be analyzed thoroughly.

Eight fundamental stages are there in an integrated manufacturing system

(1) Processing of raw materials

(2) Production of components

(3) Inspection, testing and quality control

(4) Assembly of components & sub-assemblies

(5) Testing of product/assembly for performance

(6) Packaging (an important aspect, especially in respect of consumer products)

(7) Storage in warehouse (before despatch to the retailers)

(8) Transport, delivery and distribution to end users and consumers.

These are shown in Fig. 6.3.

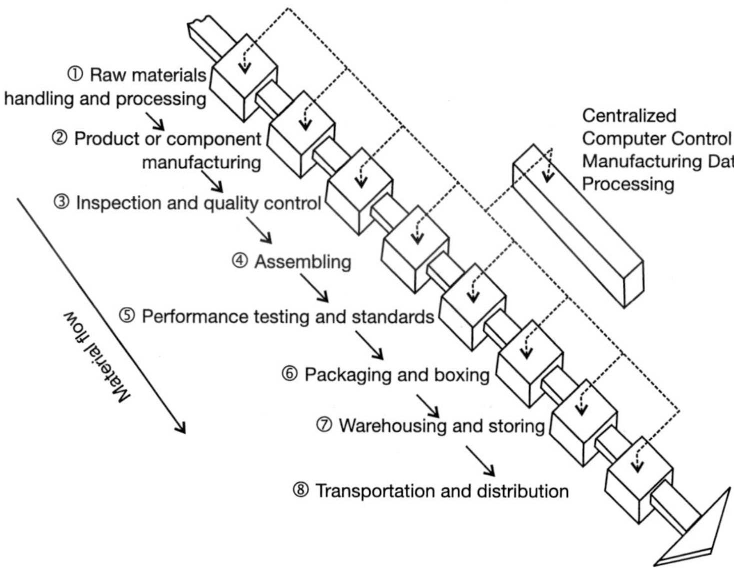

Fig. 6.3: Eight fundamental stages in automated manufacturing; groups of stages/all stages are integrated into an automated plant

Automation Planning

In the aforementioned stages, raw materials processing, parts production and assembly involves actual manufacturing. All the other functions/stages are materials handling, testing and control and transport are only auxiliary operations, but required to fulfil the complete business requirements.

What are the Advantages of Automation (as applied to Automated manufacturing)?

Computerized machines/equipment has the power and ability to perform manufacturing operations more *precisely and more efficiently*, than its human counterpart. Hence the computer aided manufacturing machines are increasingly used in industries.

Advantages of Automation:

- Reduced manufacturing lead time
- Automation results in higher production and higher productivity (i.e. higher output per hour of labour input)
- Automation is a means of reducing costs (especially the labour cost)
- Automation leads to production of high quality/precision products
- There is also reduction in in-process inventory
- Automation increases the safety of the operators, generally
- Automation helps the companies to remain competitive when considering other industries.

Automatic Manufacturing (or Automated manufacturing) involves 4 (four) elements

(1) *Automatic transfer (movement) and handling of components* between processing units or work centres. This is achieved by conveyor systems, (roller conveyors) and transfer arms. In modern plants (e.g. automobile manufacturing) *industrial robots* are used for this automatic handling (pick and place robots).

(2) *Automatic in feeding of parts to processing units* using dial feeders, hoppers (for small components), indexing tables, stock reels (for feeding of sheets to processing units).

(3) *Automatic control of processes* during the cycle of operations i.e. automatic control of predetermined timings (duration), pressures, temperatures, feeding measured quantity of powder, liquid or material. This is achieved by microprocessors/electronic controllers (temperature controllers).

Automatic Control of Process Parameters: (*Sequence of Steps involved*)

- Sensing and measuring the process parameter (e.g. temperature, humidity, etc.)
- Control of the process parameter within the limits set, and sending the appropriate signals whenever the process parameter deviates from the set values
- Processing of the signal received and taking action to control the parameter.

(4) *Automatic discharge of finished products* (*from the assembly line*).

It is done mostly by Ejectors (mostly pneumatic or hydraulically operated). The part or the product passes through counters (to know the quantity of parts produced) and falls by gravity into the bins chutes and conveyors which move the parts to the labelling packing and wrapping units.

The above elements of automated manufacturing are planned, designed and built-up suiting the product and the operations involved.

Assembly Planning

Assembly involves fitting together two or more separate components to form an assembly (completed product) or sub-assembly (a part of final product). An assembly may have two or more sub-assemblies.

Automated assembly uses mechanized and automated devices in assembly lines.

There are *three methods* used in industries to accomplish the assembly process.

- Manual single station assembly.
- Manual multiple work station assembly line.
- Automated assembly line.

1. *Manual Single Station Assembly*: Consists of a single work place or work station in which the assembly work is done on the product which is kept stationary. This is used on a product that is complex and produced in small quantities. One or more operators are employed in each workplace depending on the size of the product and the production rate required. Some examples are:

- Machine tools,
- Aircraft assembly (Fig. 6.4),
- Ships,
- Custom made industrial equipment (earth movers),
- Prototype models, and
- Automobile vehicles (cars).

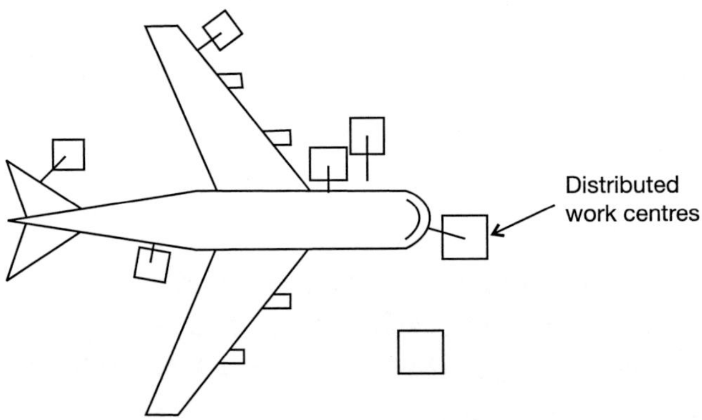

Fig. 6.4: Assembly of an Aircraft
(an example of a Manual single station Assembly)

Hand tools and power tools are used for the assembly work.

2. *Manual Multiple Station Assembly Line*: Here multiple work stations are used. The main part or a sub-assembly is kept on the assembly line, assembly is carried out and moved from

one work station to next work station along the assembly line/flow line. At each work station one or two operators perform a portion of the total assembly work by adding one or more components to the existing sub-assembly. The work is completed when a product comes off the final station.

3. *Automated Assembly Line*: Here multiple work stations are used. Automated methods (rather than human operators) are used. *Mechanized and Automated devices* are used to perform the various functions in an assembly line and to move/to transfer the assembled part (product)/sub-assembly between work stations.

If a large no. of work elements are involved, *proper balancing of the assembly lines* is required. This is to ensure that the completed sub-assemblies come simultaneously from each assembly line. Computer systems/programmes have been developed for this purpose. The *precedence* (i.e. correct sequencing) of the operations are worked out, cycle time estimated and controlled to ensure the production rate required. Production rate is expressed in completed units/hr. usually.

Computerized line balancing methods and software packages have been developed to facilitate *automated assembly planning*.

Automated manufacturing/continuous manufacturing: Continuous manufacturing is the production of goods on a flow basis at a predetermined production rate. In continuous flow manufacturing the product/sub-assembly move constantly from one operation to next operation without any storage at any point in the assembly line.

(For example, automobiles, gear box housings, home appliances (refrigerators, washing machines), food processing, and chemical products.)

Usually a materials handling equipment (conveyor system) is employed. The base frame of the product is usually the starting point and the other parts/sub-assemblies being added or mounted, as the product builds up (grows) as it moves along the assembly flow line. Final finished product is being taken off from the conveyor at the end of the line.

Fig. 6.5 shows an automated manufacturing line.

Fig. 6.5: An automated manufacturing line.
(An industrial robot doing spot welding on the automobile body)

Fig. 6.6 shows an industrial robot doing operations on automobile body shells.

Line Balancing, means an assembly line is balanced; that is to ensure that the cycle time in each work station along the line is adjusted (or speeded up) such that it is almost the same so that there is effective steady flow in the line instead of parts being accumulated between two operations (work stations) which affects the production rate.

In most cases, loading and unloading of the parts is done with the help of some material handling equipment (lifting hoist, etc.).

Fig. 6.6: Automobile body shells moving from one work centre to another through automated welding operations

Evaluative Questions

1. Distinguish between "Mechanization" and "Automation".
2. What are the distinct stages in the development of an automation system?
3. List some of the automation technologies available and are used in modern manufacturing companies.
4. What are the methods used to achieve automation in assembling final product?
5. What is line balancing? What is its importance in the context of manufacturing automation line?
6. What factors influence the manufacturing lot size?
7. What is meant by "throughput" in the context of manufacturing operations?
8. What are the benefits of "Kitting" in assembly operations?
9. What is meant by "Partial automation" in contrast to full automation?
10. What are the advantages of "Automation"?
11. Give some examples of "Mechanization".

7

Product Life Cycle and Design for Manufacturing

New Product Development Cycle

Product Development: Product development is a bridge between research work and engineering work i.e. realization of the product. The work of product development groups consists of making several product designs, testing and evaluating these designs. These testing and development is done in the market by making a few numbers, getting feedback from consumers and also testing under laboratory conditions. One best design is chosen and further developed.

The manufacturing processes are also identified for making in large quantities; the capability of the machines and facilities available are also examined. Any minor changes/design modifications are also done in consultation with designers. Cost information/estimating is also done.

Product Development Cycle: Seven steps are involved in new product development:

(1) *Obtaining ideas* from research group, customers and designers about the potential of the new product.

(2) *Screening of ideas* by top management, sales and marketing staff based on their prior experience.

(3) *Technical evaluation* consisting of method of production, cost of production, and performance of the product.

(4) *Market evaluation* acceptability in the market and at what price?
(5) *Final decision* by top management and finance/budget availability.
(6) *Getting into production* any new equipment, tooling, inspection gauges required, materials/parts to be ordered?
(7) *Introduction to the market* by sales and advertising department, advertisement media, packaging, servicing and channels of distribution.

Product Life Cycle (as considered from product design/ engineering point of view) is shown in Fig. 7.1.

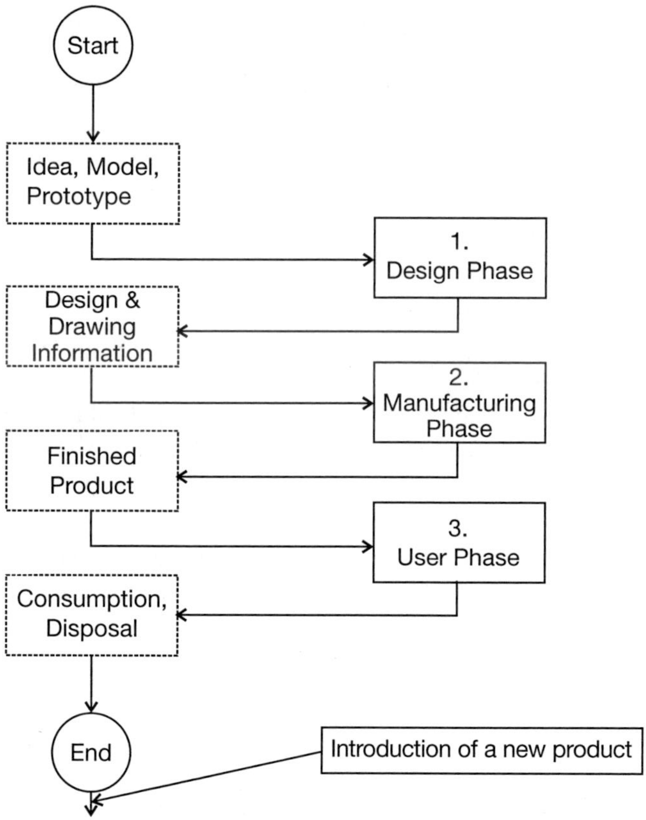

Fig. 7.1: Life cycle of manufactured product. Consists of three distinct phases (from Designers point of view)

Product Semi-Realization

This refers to product design partially carried out to meet the immediate requirement of the management as soon as an ideation occurs regarding the need to develop a product (i.e. *felt needs*). This helps the top management take a decision whether it is worth pursuing the task of going ahead with the detailed design. For semi-realization conceptual design during the design phase is useful.

> There are *three distinct phases in the Design stage* when new products are designed and developed:
>
> 1. Conceptual design
> 2. Preliminary design
> 3. Detailed design

When all the three design phases are completed *product realization* (full realization) in the physical form occurs.

In the conceptual design, Geometry and material details are not known except in the case of parts which dictate the essential product functioning. The product designer gets an idea of the assembly, sub-assembly and component parts required.

In the conceptual design, the most important factor considered is the functionality and reliability of the product. *Semi-realization of the product occurs during conceptual design.*

In the preliminary design, the types and details of materials are identified and the product and parts are completely dimensioned. Dimensions, its limits, fits and tolerances (and allowances) and surface finishes are decided now.

Production processes used for the manufacturing of the product (manufacturability) are determined at this stage.

Decision is taken as to what extent standard parts can be incorporated in the final assembly.

In the detailed design, complete information of the product is known and product design is completed. Production processes and the process parameters are also decided.

One or two prototypes are made to know how the product will look like (visualization) and its performance with respect to its functional requirements.

After completing all the three phases full realization of the product occur.

Next, industrial engineering and shop floor practices such as:

- Rote sheets/process sheets
- Production details (materials required, lot sizes)
- PPC (Production Planning & Control) procedures
- Scheduling

are finalized, since the new product manufacturing is to be taken up in the context of manufacturing lines already engaged in the production of old products.

Cost Estimation calculations can also be done simultaneously as the product design goes through these three stages to have an idea of the *cost of the final product* (*Target cost*).

Design for Economics (i.e. design for economical production)

Evaluation of Design for Manufacturing Cost

First, checking whether the part/product designed will meet the functional requirements?

Then, the part designed is evaluated and critically examined so that manufacturing cost is minimum.

The following are the *criteria adopted* for evaluating the design for manufacturing cost.

Manufacturability

- Whether the "*Manufacturability*" or *producibility* requirements have been considered? (i.e. product design, material chosen and processing method have been considered in an integrated manner in the initial stages itself.) This will result in minimum design changes.
- Whether the part is designed so that the manufacturing is possible using the existing facilities (i.e. machine tools, equipment).

Design for Manufacturing (DFM)

- Check whether design facilitates easier machining, testing and assembly? (Design hints are available in handbooks for designing the part for easier machining, hardening (heat treatment), testing and assembly).

– Check whether the *types of holes* and other parts features specified are *standardized* so that there is a *minimum variety of tooling requirements*.

– Check whether limits, fits and tolerances specified are within the process capabilities of equipment and machine tools available in the plant. (Designers have the tendency of specifying very close tolerances and finish requirements which may not be really required for the satisfactory functioning of the part.)

e.g. • Grinding is a costlier operation when compared with machining process.

• Fine finishers are not required for external surfaces of parts.

– Check whether the locating features, fixing, clamping and insertion of parts during assembly is easier and there is no difficulty in the assembly and the assembly is foolproof.

Modular Design

– Check whether the concept of "Modular design" is adopted so that the "Modules"/pre-assembled kits can be fitted in the assembly quickly instead of resorting to time-consuming assembly methods. (Modular design also facilitates quicker replacements of parts during servicing/ maintenance work.)

Standardization

– Check whether standard parts/materials have been specified which are easier to procure and are cost effective. This will also help in eliminating a wide variety of materials/components which will result in minimum manufacturing cost.

All the above will result in less cycle time, quicker production (higher production rate) and hence less cost and manufacturing economics.

Implications of the Design on the Manufacturing Cost

A product Designer must understand that whatever he designs will need to be justified in terms of cost of producing the items. He should understand the implications of the design features of the parts. He should at this stage consider how a part designed can be made economically, without sacrificing any of the functional requirement and safety aspects.

There are design tips available as to how one can design the part for economical casting, for economical machining and inspection and for economical and ease of assembling. These are collectively referred to as *DFM* (*Design for manufacturing*) **and** *DFMA* (*Design for manufacturing and assembly*) for which design softwares are available.

(1) *Examples of Design for Manufacturing* (Design for Machining) guide lines are:

- Design for minimum no. of parts.
- Minimize part variations.
- Avoid separate fasteners (use standard sizes of fasteners, nuts, bolts, washers).
- Avoid drilling on thin walls and drilling on inclined surfaces.
- Avoid blind holes (through holes are easy to machine/drill when compared with drilling of blind holes).
- Keep centre distances between holes, so that there is minimum interference when holes are drilled.

(2) *Design for Measurements* and guidelines for design of *castings, and heat treatment of parts*, are also available which a product designer should be aware of, which should be considered in the initial stages of product design.

(3) *Examples of Guidelines for Design for Assembly* (DFA)

- Use mechanical feeding devices such as feeders, hoppers, chutes for placing of parts for assembly.
- Minimize handling of parts to a minimum so that assembly is easier (and not frequent turning of jobs during assembly).
- Use fool-proof methods for assembly (i.e. parts are not kept upside down or assembled on the reverse side).
- Design the assembly fixtures such that they can be clamped using hydraulic/pneumatic power. This will minimize the fatigue of the assembly operators.

- Whether Robots can be used for assembly? (for placing the parts, clamping and dropping the assemblies in the bin.)

(4) *Design for Maintenance*

- Products are designed such that disassembly is made easier and quicker; defective/broken part removed and new part inserted and assembled back. This will be useful in servicing and maintenance.
- During maintenance whether standard tools can be used instead of specialized assembly tools. (e.g. bearing pullers).
- Ensure that assembly procedure follows a logical sequence.

(5) Disposability of the parts (especially for home appliances and consumer products and plastics after their service life); whether recycling of materials used in the product is possible?

Concurrent Engineering considers all these factors in the initial stages of design cycle itself so that products are designed leading to many advantages.

Evaluative Questions

1. What are the distinct phases or stages in the life cycle of a product?
2. What are the uses of product life cycle?
3. At what stage new products are designed, developed and introduced in the market?
4. Why is a systematic evaluation of the design of a product or part should be made which respect to manufacturing cost?
5. What is meant by "manufacturability" or "producibility" of a part?
6. What is meant by DFM (Design for Manufacturing)?
7. Write a note on "standardization" w.r.t. the product design.
8. Give some examples of (a) design for machining (b) design for assembly (DFA) and disassembly (c) design for maintainability.
9. What are the three distinct stages involved in the new product design?
10. What is meant by *modular* design?

8

Manufacturing Cost

One of the significant factors in manufacturing cost is the *product volume* (i.e. quantity or no. of pieces produced).

Large product volumes require high production rates. High production rates are possible by using special purpose machines/transfer machines and specialized tools and equipment. Manufacturing cost per piece gets considerably reduced if components are mass produced by *mass production techniques* involving use of SPM (special purpose machines) and dedicated machinery.

Small batch production (*small lot sizes*) is done by GPM (General Purpose Machines), e.g. lathes, milling machines and conventional NC machines. Manufacturing cost per piece is relatively more in case of small volume production. This is illustrated in Fig. 8.1.

Economical Lot Size

Economical lot size (i.e. *no. of units processed as a batch*) *depends on* cost of raw material, inventory cost, set up cost, working capital required (i.e. capital required for day-to-day running of the plant), and cost of storage of the raw material (carrying charges) and the % interest rate.

Batch size is expressed in no. of units.

Production cost: Part design, material choice, processing conditions, type of equipment used and lot size all contribute to the ultimate product cost.

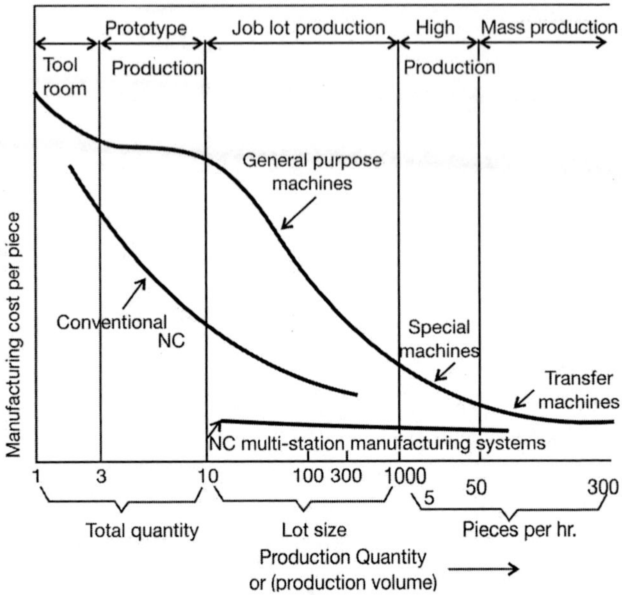

Fig. 8.1: Manufacturing cost per piece as a function of batch size for different manufacturing systems. Mass production and specialized automated production methods result in high production rates and reduced unit manufacturing cost

Manufacturing Cost and Importance of Economical Lit Size

The larger the *lot size* (*i.e. batch size*), the lower is the cost per unit to set up the machines and to prepare the process plan and for production planning and control.

At the same time larger the lot size, the greater is the amount of raw material consumed and the labour cost. This increases the amount of inventory stored for the bigger lot size. This increases the total cost of production. Hence there is an optimum/ economical lot size.

Economic lot size is given by a formula:

$$L = 5\sqrt{\frac{MS}{KC}}$$

Where L : Lot size, no. of pieces

M : Pieces produced per month

K : Carrying charge factor, % (from 5-30%), i.e. cost of storage of raw material, expressed as a % of material cost used for one lot

C : Material cost per piece (₹)

S : Set-up cost per lot (₹)

(S : Set-up cost per lot depends on set-up time and labour cost of set-up engineer/operator who sets up the machine). The above formula is a guide only (approximation only), but can be used as a starting point for estimating costs.

Problem

A company produces 2000 pieces/month in its assembly department. The cost of each piece is ₹ 25 and the cost of each set up is ₹ 50 per lot. If a carrying charge factor of 30% is used calculate the economic lot size.

Solution

The economic lot size is $L = 5\sqrt{\dfrac{MS}{KC}} = 578 \text{ pieces (say 600 pieces)}$

M = 2000 pieces/month

S = ₹ 50 per lot

C = ₹ 25 per piece

K = 30% (carrying charge factor)

Manufacturing Cost (Fixed Cost and Variable Cost)

Manufacturing cost can be divided into *two major categories*: *Fixed cost and variable cost.* These are usually expressed as *Annual Costs.*

Fixed Cost	Variable Cost
It is at a constant level, independent of the level of production (i.e. volume of production or production output) • Cost of factory building • Cost of equipment/machinery (machine hour cost) • Taxes • Insurance on equipment/ factory building	It is proportional to the production output i.e. depends on volume of production. • Direct material cost • Direct labour cost • Cost of energy consumed for production of components

When we combine both the fixed cost and the variable cost, we get the total cost of manufacturing. Total cost depends on output volume (annual output) of the products (units). Total manufacturing cost increases with the output level.

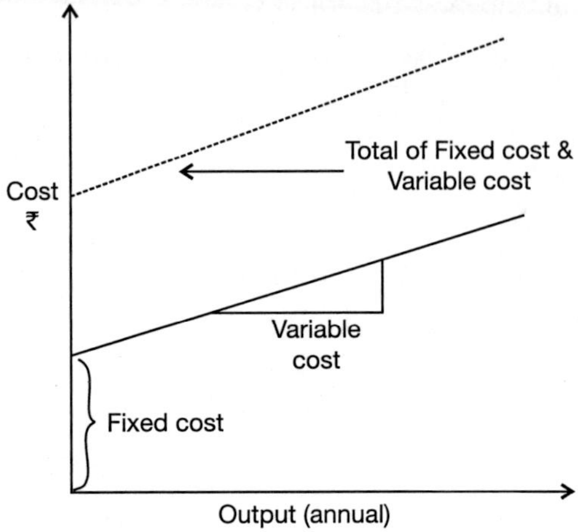

Fig. 8.2: Fixed cost & variable cost as a function of product output (units, annual), total manufacturing cost = fixed cost + variable cost

In addition, there are overhead charges/overhead cost. These are costs associated with running a firm. There are two types of overhead charges, *Factory overhead charges and Corporate overhead charges.*

Factory overhead charges (Factory Expenses)	Administrative/Corporate overhead charges
• Cost of repair and maintenance of equipment and machinery • Factory security staff, transport staff salary, maintenance staff • Cost of energy consumed • Cost of spare parts for maintenance of machines	• Salary of executives • Administration cost • Office building, rent • Salary of design engineers, R&D, sales, accounts and finance staff • Cost of Air-conditioning

Methods Available to Calculate the Cost of Manufacturing

There are many methods available to calculate the cost of manufacturing a product by a company. The cost is defined as the expenses charged to an account for items produced, the cost of distribution of items sold and the cost of rendering the services.

The cost of manufacturing refers to:

(a) Direct costs, and

(b) Overhead costs.

Direct costs are those that enter directly into the manufacturing of the product.

These are further divided into:

- direct material cost (cost of raw material that becomes part of the final product).
- direct labour cost (cost of machine operators' wages that are paid to the labour/operator).

Overhead costs (overhead charges). These are cost incurred for the activities that do not enter directly to manufacturing of a product.

These are further divided into:

- indirect material cost (cost of materials used for building jigs, fixtures, gauges, and standard parts such as rivets, spares, bolts and nuts).
- indirect labour cost (inspection cost, set up cost, maintenance of machines and tools).
- fixed costs (insurance cost of building, equipment, taxes, rent and depreciation).

Overhead costs arise from peripheral activities that have no relationship to direct production output, but which affect the production costs. Management (cost of salary of managers and executives), engineering staff, office, plant (equipment), indirect labour cost and cost of standard parts and spares constitute overhead charges. These are costs that support the enterprise without which there could be no efficient production.

Total Cost Function: Total cost depends on annual output expressed in units. In addition, it shows of the units. This is proportional to the output (units sold).

These two lines i.e. total cost function line and revenue line (income from sales) intersect at *point A*. This point is known as *Break-even point*.

(1) *Break-even chart is used for profit analysis*: If the annual output (units sold) is greater than the break even point 'A' there will be *profit*. If the annual output (or units sold) is less than the break even point, there will be *loss*. (Fig. 8.3)

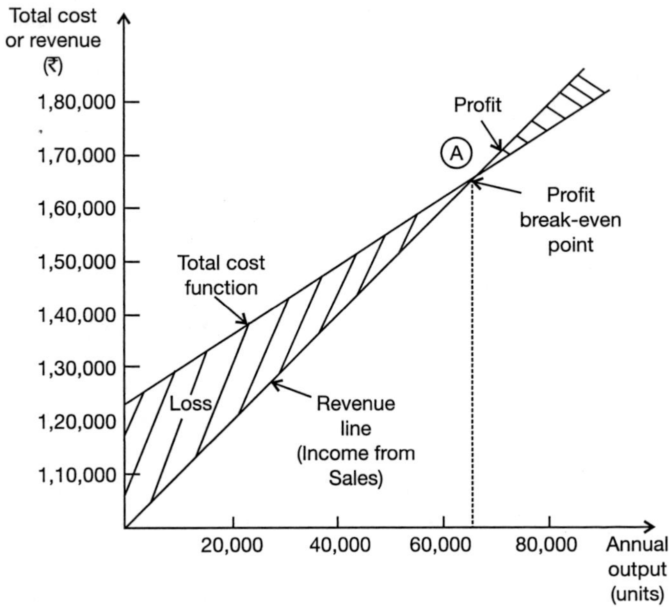

Fig. 8.3: Break-even chart used for profit analysis. A is the Break-even point

(2) *Break-even chart is also used for comparison of manufacturing costs by different production methods*

This is shown in Fig. 8.4.

The cost function for a manual method of production and for an automated method of production is shown in this chart. In automated production method, fixed cost is more due to installation of costly machinery, but variable cost is less due to less machining time. In ordinary manual method, fixed cost is less, but variable cost is more since machining time/manufacturing time is more. At point B (known as break-even point), total cost function is same (equal) for both the methods of production. For output levels beyond point B, automated method of production is economical; for output levels less than point B, manual method is economical.

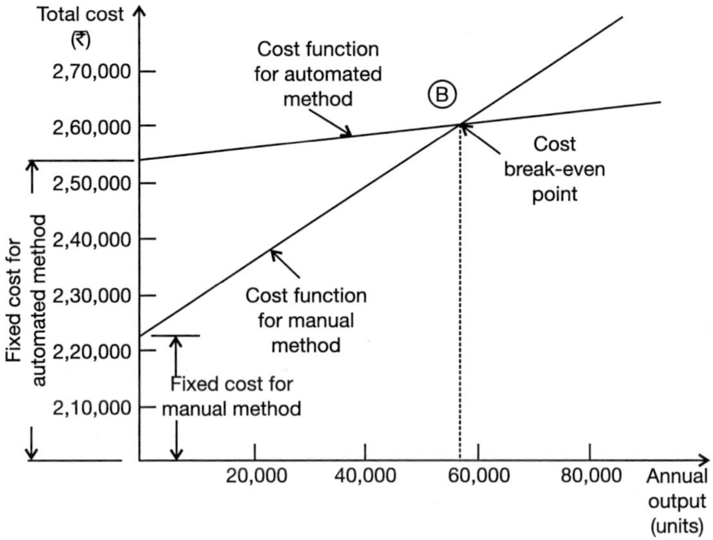

Fig. 8.4: Break-even chart for comparison of manufacturing cost by manual method and automated production method (B) is the break-even point

Actual Cost Vs. Standard Cost

There are two major methods used for calculating the costs based on (i) actual cost of production, and (ii) standard costing method.

Actual cost of production refers to cost of material actually consumed and labour cost. This cost is calculated after production starts.

Standard cost method is based on standards determined before actual production starts. Standard costs are worked out by companies based on their past production data, standard prices and tables of labour cost (time standards). Companies are expected to calculate and maintain the past records for this purpose (i.e. to arrive at standard costs), for estimation of costs of new products.

Cost Control

Cost control is the prevention of waste in all respects in a manufacturing environment. *Standard costs* are developed based on the historical production records maintained by the company over the past years.

Cost control is the procedure whereby *actual costs* incurred are *compared against standard cost* so that unnecessary costs (considered as waste) are identified and measures are taken to eliminate them or correct them. This is known as cost control. (e.g. unnecessary costs are incurred due to *poor design of products*, unproductive time/labour, wastage of material and other resources, scrap, etc.)

Cost Reduction. It is the result of improvements made by better product design and good manufacturing practices. Ways by which wastage occurs are identified and measures taken to eliminate them.

Cost reduction is a continuous *activity requiring* constant appraisal of the various activities of the company (and not just production processes) by *improving productivity* in all the activities of the organization.

Cost control leads to cost reduction and both are highly inter-related.

Everyone in the organization is responsible for cost control and cost reduction. The suggestions given by operations staff and engineers should be considered by the management.

Evaluation of Design for Manufacturing Cost

In order for a manufacturing organization to remain competitive it must make its products at the minimum cost consistent with the required function and quality of the product. The company will then make the maximum profit possible which will ensure continuing financial health of the organization. Manufacturing engineers and design engineers must always be cost conscious. Product design made by the design engineers must be objectively evaluated for manufacturing cost.

Cost evaluation, cost calculations or cost estimating is an important aspect in the *design stage of a product.*

It has been established that at design stage itself about 80% of the final product cost has been committed. Hence product designers should be more *cost conscious* when they are designing the product or part. Significant cost savings can be realized when the designers are able to evaluate their design on a cost-effective basis, and hence unnecessary costs can also be eliminated. This is essential for *long term success of any concurrent engineering project.*

There are three distinct stages in the design of a product:

(1) Conceptual design,

(2) Preliminary design, and

(3) Detailed design/final product design.

(1) *In conceptual design stage*, the functionality of the product is ensured. Weight and volume of material is considered consistent with the product performance required. Geometry/ dimensions of the part are not known. Material used is also not known. Manufacturing process(es) are also not yet known. Consideration is given to keep the assemblies, sub-assemblies and no. of parts required to a minimum.

In conceptual design stage *accuracy of cost estimation/cost evaluation is expected to be about 50%.*

(2) *In the preliminary design stage*, the material and any specific requirement of the material is known. *Cost evaluation at this stage is 30% accurate.* Existing cost data base and past information concerning the cost is used. Tolerances and surface finish requirements are chosen so that cost is minimum consistent with the part/product's performance.

It is during the conceptual design and preliminary design stage, designer can bring about maximum cost reduction.

(3) *In the detailed design stage,* all the information about the product is known as a result of performance of the prototype and the design is complete. Material and its specification are known. Final dimensions, tolerances and finish requirements are finalized. Manufacturing process is known and the process parameters are also established. A detailed cost estimate at this stage will be to an accuracy of 15%.

This approach for *evaluation of design for manufacturing cost* can improve the competitiveness of the company to a great extent since the company can fix the *cost of their product*(s) on a *realistic basis*, and the market share set is realized.

Life Cycle Costing and Terotechnology

"Terotechnology" is defined as a combination of management, financial, engineering and technology and technological practices, applied to physical assets (machinery and equipment) in relation to *economic life costs and the benefits* derived from it, (*throughout the life of the product*).

The *benefits of terotechnology* are:

- Overall cost reduction in all spheres of manufacturing activities.
- Reduced breakdown costs (cost of stoppages in production and losses due to production delays).
- Lower material cost (less scrap).
- Improved quality and efficiency of manufacturing operations.

Importance of Predictive Maintenance and Preventive Maintenance

Whenever the machinery or equipment breaks down, it has to be repaired, defective parts replaced and servicing is done to restore the machine to its original shape and condition. In practice, the cost of maintenance is found to be higher than the original cost of the machine. No. of breakdowns and production delays and losses in production can be controlled by predictive and preventive maintenance procedures and schedules.

> **Life Cycle Costing**
>
> The cost of equipment, operating cost, and maintenance and servicing cost of the machinery and equipment during their *entire productive life* is known as life-cost costing.

Life Cycle Costing consists of two important components

- Visible costs, and
- Hidden costs.

Here an analogy of an Iceberg is used.

These two costs are shown in Fig. 8.5.

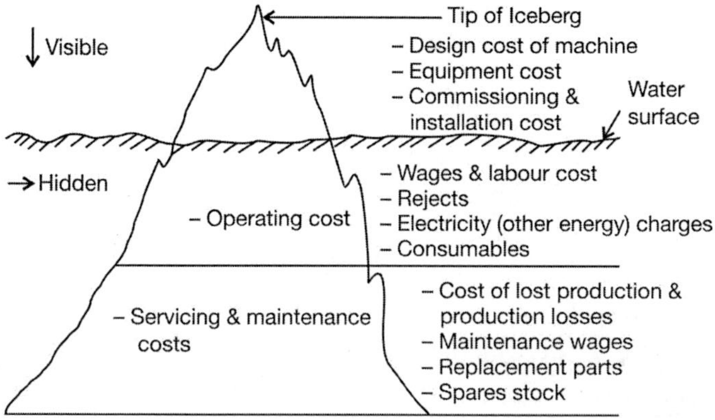

Fig. 8.5: Life cycle cost showing *Visible cost and Hidden cost*. (Analogy of Iceberg is used). Tip of the Iceberg is visible cost; Hidden cost consists of operating cost and maintenance cost. Life cycle costing takes into consideration all the cost incurred during the productive life of the equipment and the benefits derived from it. (cost-benefit analysis)

The following specific cost factors should be considered in the *Life cycle costing and benefit analysis.*

- In *visible cost*, design cost is incurred especially if new design of equipment or product is required.
- In respect operating cost: generally *labour cost and energy* cost for operation of the equipment *increases* as the years pass by.
- In respect of *servicing* and *maintenance* cost:
  Breakdown maintenance involves dismantling of equipment, replacing defective parts and assembling back. This involves considerable amount of time. Here concept of *modular design* is useful wherein defective sub-assembly is quickly replaced by a standard module. This could be considered during the design of the product.

The hidden cost is more when compared with visible cost.

Deciding *when to replace the existing equipment*: This depends on:

- Extent of deterioration in the performance of the equipment/machinery, due to continuous operation of the machine.
- Reliability of the product/equipment at any point in the life span of the equipment.
- Ease of servicing, repairing the defective part and maintenance.
- Ease of operation of the equipment.
- Overall *cost-benefit* economics (i.e. in essence getting the *higher benefits for least cost*).

Concurrent Engineering should take these aspects into consideration when designing and developing new product. This is the importance of *Life Cycle costing and Terotechnology.*

Life Cycle Engineering

The philosophy of life cycle engineering is that the entire life of the product should be considered in its original design.

In addition to the basic requirement of functionality, performance, reliability, and safety maintainability should also be addressed at the design concept stage.

Close integration of all aspects of manufacturing in the design stage is required.

Evaluative Questions

1. Give some examples of
   (a) Fixed cost
   (b) Variable cost
2. What are *Factory overhead charges*? How will you minimize them?
3. What is meant by Break-even point? What is its significance w.r.t. manufacturing cost?
4. With the help of graphs explain how break-even point chart can be used in *Profit analysis*.

5. How is break-even chart can be used for comparison of two processing methods?
6. How will you exercise cost control?
7. Explain how *cost reduction* can be made?
8. How cost evaluation can be made in different stages of product design?
9. What is the accuracy of *cost estimation* possible during *conceptual design stage*?
10. What is meant by *Life Cycle Costing*?
11. Define
    (a) Direct cost
    (b) Overhead charges
    (c) Fixed costs
12. Explain the difference between the *actual cost* and the *standard cost* methods used to calculate the manufacturing cost.
13. Define *Economic Lot size* and what is its relevance to concurrent Engineering?
14. Choose the best alternative.

    The difference between "Actual cost" and "standard cost" is that:

    (a) Actual cost is always greater than the standard cost.
    (b) Actual cost is always less than the standard cost.
    (c) Standard cost is calculated before the production commences.
    (d) Standard cost is calculated after the commencement of production.

    Key: (c)
15. What is *Terotechnology*?
16. In "Iceberg analogy", *Hidden cost* is much higher than *visible cost*. Explain this statement.

9

Concurrent Mechanical Design and Product Realization

Geometric Modelling of a solid object (or any component part) is the first step in product development. Solids or objects, however complex or complicated as they may be can be visualized as consisting of many *elemental solids*. These elemental solids are also known as *Primitives*. (Fig. 9.1)

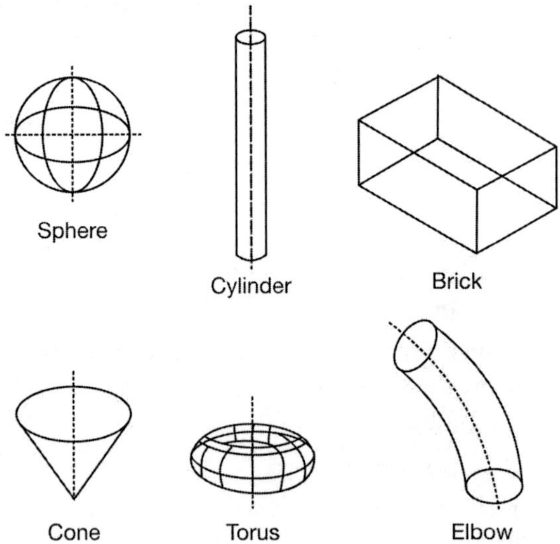

Fig. 9.1: Elemental solids or primitives used for constructing a solid (solid modelling), for analysis of stresses and strains

The primitives or elemental solid objects are in the form of:
- Spheres
- Cylinders or rods
- Cubes or rectangular blocks
- Cones
- Torus (annular rings), and
- Elbows

Solid objects are constructed/built-up by manipulation of these primitives (elemental geometries) into desired solid objects.

Different manipulations of elements possible are:
- Combination (also known as union), i.e. additives of elemental objects
- Substraction (removal of certain elements from a solid block) e.g. drilled hole can be considered as a substraction of a cylinder from a solid object
- Turn it upside down; reverse it; inside out
- Intersection (i.e. of a cylinder into another cylinder).

CAD Softwares, are available for this manipulation. These manipulations enable the CAD engineer to quickly visualize the shape, form and size of the object/component on the display screen of the computer. Creative designs can be obtained as a result of these manipulations.

The no. and minimum size of the elements (elemental objects) and geometry depends on the resolution and accuracy and the computing capacity available.

Decomposition

It is the reverse of the building up of the solid geometry. A solid object can be decomposed or split into its constituent elements or a finite no. of elements. This helps in the analysis of stresses, strains (deformations) to which a structure is subjected to static and dynamic loading conditions; and failure analysis can be made, simulated and depicted on the computer screen. This technique (i.e. decomposition) is useful in concurrent engineering design and any design modifications can be carried out easily and quickly.

The above is known as *hierarchical spatial* (i.e. *space*) decomposition.

Product Realization

A new product is realized through a series of manufacturing activities. The *information flow* between various stages of manufacturing activities is shown in Fig. 9.2.

The *Taxonomy* of different terms associated with (new) product realization in a manufacturing environment is explained below:

- *Product Design.* This takes into consideration the functional requirements of the product. This is done by the company's designers/design group. Main assembly, sub-assemblies and individual components, the materials required, their details/part features (dimensions, shapes, etc.) are clearly specified in the manufacturing/production drawings. CAD softwares are extensively used, now-a-days for product design.
- *Process Planning.* Process planning specifies the sequencing of manufacturing processes, from raw material to final product assembly. This is done by manufacturing engineers/process planners.

At this stage, *integrating the product design* (CAD) and *Process Design* (CAM) is done. This is the first and important stage in information flow. Traditionally, the two functions are considered as separate activities. Concurrent engineering ensures that this integration, viz. *Product design and process planning* takes place first in the hierarchy of product realization.

- *Manufacturability* of a particular design is mutually discussed between CAD engineers and manufacturing engineers and any modification required in the design, without sacrificing any functional requirement, is approved by the product design team. It is also ensured that manufacturing cost is kept to a minimum.

- *Operations Programming*. This primarily refers to CNC machines programming. CNC machine operations programming are done by part-programmers using the standard Alpha-Numeric Codes.
- *Verification and Validation* of the part-programme developed by the CNC machine engineer is carried out by running a trial version of the process, checked, verified, validated and approved. *Cycle time* taken for manufacturing the part is noted for any reference later.
- *Scheduling*. This is done by studying the production scheduling practice followed by the company, since the making of the new product is to be done along with the existing products using the same manufacturing line/ facilities.
- *Execution*. This refers to releasing of final manufacturing order to the production shop for producing the first batch of the new product.

Fig. 9.2 shows how *information flow* takes place in product realization whenever new products are conceived, produced and introduced in the market.

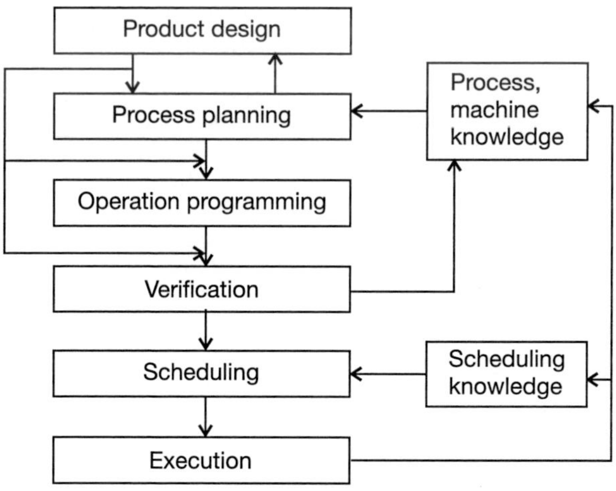

Fig. 9.2: Product realization and the corresponding information flow from product design to actual production of the part/product

Product Taxonomy Terms

Taxonomy refers to the method or way of classifying things. Various terms related to product are described below:

- *Product Attributes*: Involves listing of the characteristics of attributes of a product and identify the possible changes which might bring about an improvement. The attributes, in general are size, shape, colour, capacity/ volume, weight and other specification of the product.
- *Product Design Brief*: It is a document or formal outline with logically related headings that explains major points of a product desired. The brief/document provides basic instructions and background information required by the industrial design to design a new product.
- *Product Commerce/Commercialization*: The activities related to making the product(s) available for sales and marketing. It includes developing a product by manufacturing in sufficient quantities, economically and in a cost-effective manner, and making it available for distribution and consumption.
- *Product Designer/Industrial Designer*: A designer/ specialist involved with visual communication, and understanding of arts, aesthetic qualities and ideas. (Colour combinations). In product development, he/ she plans the layout, form, shape and structure of new products. He also decides whether a particular coating or finish is required to improve the aesthetic quality and appearance.

Convenience in the use and maintenance of the product is an important aspect considered by the Industrial Designer.

What is a Good Design?

Good design is finding the right solution, defining it, and then creating an effective, economic and harmonious solution. Good design is not only external styling; it comes from within.

Product Engineering

Product engineering is mainly concerned with product design. It also includes R&D. Product engineering is designing of products for commercial use and applications.

The elements of product engineering are:

- Design of component parts.
- Design for manufacturing (DFM)/manufacturability.
- Design for maintainability.
- Design for Disposability.

Product Research: This is an effort required to ensure continued prosperity of the company. Product research results in developing new and different products, new uses/application of existing products, or utilization of a by product. The stimulation for product research can come from different people, especially from customers/users, product design group and from industrial consultants. Product research also concerns with expanding the product life. Ultimately all this should result in development of a useful and saleable product.

Product Development: The complete process or network of activities by which companies bring out new products to the market, starting from the conception of new product ideas upto the point of launching it commercially in the market (commercialization activities in the market is also considered as part of product development process). In essence, product development is a bridge between research work and engineering work.

Product Development Life Cycle: Start with ideation of the product, its development, growth of the product in the market, maturity, stagnation and coming up with a new product with new features. Sales and profits are always kept in mind in the product life cycle.

Design Review: A review of designs for manufacturability (or producibility) to check for compliance to workable design criteria. The cross functional group interacts and vital information about the design is exchanged. The design review team identifies the problem, determines the course of action and corrects the problem areas.

Redesign: As a result of decisions made at design reviews the details of design are changed, prototypes are developed and tested.

There are two categories of *Redesigns:*

(i) Design modification is done so as to improve the product performance, and/or

(ii) Enhancing the capacity of the products, to improve the appearance and aesthetic appeal of the product to keep it competitive.

Product Evaluation: It is the systematic consideration and evaluation of new product ideas. The process can be either subjective or can involve objective measurement against predetermined criteria. The main purpose of evaluation (or sometimes referred to as "*Screening*") is to select *potentially profitable new product ideas*.

New Product Rating: It is a *conceptual model* (an algebraic expression), mathematical model showing the relationship between life of a product, profits growth, years and total cost of development (initial cost and recurring cost). Used to compare the *relative worth of various product ideas*, to enable top management to take reasoned decisions.

Product Realization Taxonomy

Different terms associated with *New Product Development*:

- Product attributes
- Product design brief
- Product commerce/commercialization
- Industrial designer
- Product engineering
- Product research
- Design review
- Redesign
- Product evaluation
- New product rating
- Product patenting

Negotiation in Concurrent Engineering Design

Negotiation is done through discussion with the concerned stakeholders. Negotiation in concurrent engineering design occurs on four platforms.

(1) Negotiation of product design team with manufacturing engineers (process design group),

(2) With inspection staff and quality control engineers,

(3) With purchase department, and

(4) With the vendors and part suppliers.

(1) Negotiation with Process/Manufacturing Engineers: Manufacturing engineers can express their difficulties and their concerns in manufacturability/feasibility of making the product with the available facilities, machine tools and other equipment. Significant concern will be whether process capabilities of machine tools are within the tolerance specified, accuracy and surface finish requirements as specified in the manufacturing drawings.

Manufacturing engineers may request the product design team whether the tolerance requirements can be relaxed. Product design team may consider their request to the extent that functional and safety requirements are not sacrificed.

If any new production facility, equipment is required to meet the product design specific requirements, this may be recorded and management informed to provide funds to procure the new equipment.

(2) Negotiation with Inspection and Quality Control Engineers: This ensures whether the desired inspection and quality control procedures and equipment will be set in place and quality control engineers are made aware of the accuracy specifications and measurements mentioned in the manufacturing drawings.

Any special inspection equipment (such as CMM, coordinate measuring machine or Gauges) required are indicated and recommendations made to procure them.

(3) Negotiation with Purchase Department: This is to ensure that the specified material with the strength and other characteristics are available in the market. Suitability of any alternative material can also suggested by the purchase dept. for consideration by design team.

Purchase dept. can also recommend whether blanks or slugs required can be in the form of forged, cast, welded part (weldment) or need to be machined only from raw bar stock.

(4) Negotiation with Vendors/Parts Supplier: Product design team and purchase dept. staff can negotiate with the vendors to ensure that the parts supplied are as per the accuracy specifications mentioned and can be made available in the economical lot sizes specified. They will also ensure that the vendor has the required equipment to make these parts and will supply as per the time schedules (timely supply) worked out.

The concept of Modules/kits in the pre-assembled form (as a kitting) can be introduced to the vendor and whether he will be able to supply them to the company. Such pre-assembled kits/modules can be mounted on the assembly lines directly thus reducing the inventory to a minimum. Timely supply of kits will also promote JIT production in the main plant.

Negotiation w.r.t. cost of the parts supplied can also be done so as to avoid any misunderstanding in the future.

Evaluative Questions

1. What is meant by *decomposition* (of a product) in concurrent design?
2. Name some *primitives* which are used in the development of a solid.
3. What are the *different manipulations* that can be employed for building up of a solid/project?
4. Explain the various *terms* related to *product realization taxonomy*.
5. What is meant by '*Manufacturability*' in the context of a new product development?
6. What are the different *attributes* related to a product?
7. What is the role of an *industrial designer* in the development of a new product?

8. What is meant by *negotiation* in concurrent engineering design?
9. What is the importance of negotiation (by discussion) with vendors?
10. Explain the concept of *Modules* (*modular design*) and *Kitting* in the design of new product and its assembly.
11. What is a "Good design"?
12. What is meant by "Product Research"?

10

Plan for Project Management on New Product Development

Intense competition amongst the industries to execute and complete the project at an increasingly rapid pace, forces the project team to adopt project management techniques. Long term complex projects need proper project planning.

Project management may be considered as a *time compression technique* and at the same time enables maintaining of quality and avoids time over-run and cost over-run.

Project Management and New Product Development

New product development is often conceived as a project management task, to plan and execute the task in a cost-effective way and at the same time keeping the time frame in mind.

Project management involves a *systems approach to planning, scheduling and controlling the various activities* connected with new product development. To enable this to happen suitable/ right *information system* and related softwares linking up the various technological processes and activities are to be built-up and put in place. Sharing such information system will also enable geographically distributed teams to work *collaboratively* in product development.

Planning a Design Project

Planning for the development of a new product is perceived as a *Design Project to be planned and executed*. For this certain procedures and steps have to be followed as shown in Fig. 10.1.

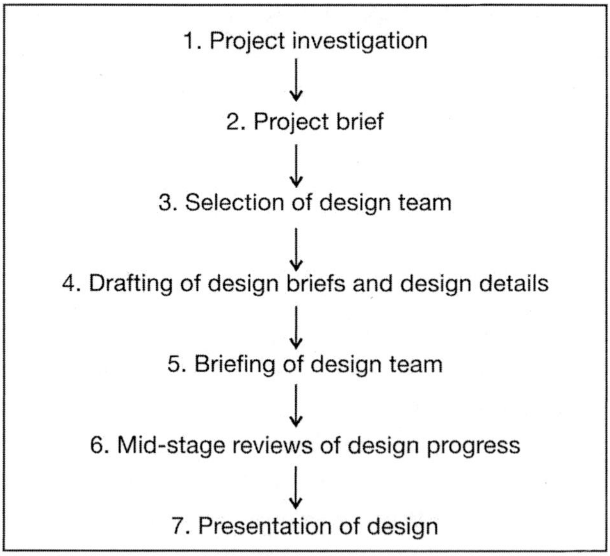

Fig. 10.1: Steps to be followed considering the development of a new product as a project to be planned and executed

1. Project Investigation: Investigate the potential and the feasibility of the project in consultation with the top management and executives.

2. Project Brief (i.e. briefing about the project): Discussing with sales, and marketing staff to ascertain the market potential of the new product and for whom the product is intended, i.e. potential customers?

3. Selection of Design Team: Design team to consist of engineering product designers and industrial designers.

4. Drafting of Design briefs and Design details

- Ensure that the brand image of the existing products are retained and extended to new product(s).
- Check whether similar products are sold by competitors.
- Consult the industrial designer for styling, form, shape, aesthetic appeal, appearance and finish expected.

5. Briefing of Design Team: Completed design details are presented to the top management and executives/managers.

6. Mid-stage Reviews of Design Progress: Establish dead lines/mile stones for monitoring of the project and for corrective action, if any, required.

7. Presentation of Design: Details of performance of the mock-up or prototype is presented and *concurrence* obtained for executing the project.

Providing Budget

For planning and executing the project:

- Design cost (cost of acquisition of design and license).
- Any special material to be procured (from abroad?).
- Tooling requirements for mass production.
- Whether existing production facilities are sufficient, and capable of manufacturing the new products.
- Facilities for prototype production and testing.
- Skill level of operators; any special training required.
- Cost of acquisition of special CAD systems and softwares for design team; and drafting.

Carry-out final review of the project and execute the same.

Processes and Steps Involved in Project Management

- Clearly identify the task to be completed.
- Determine all the activities to be performed for completion of the project/task.
- A sequencing order must be determined (*technological sequencing*).
- Build-up an *information system* for executing he project.
- Identify the well defined *starting point* (i.e. initiation of the project) and the *end point* (i.e. completion of the project).
- Estimate the time and cost required for each activity, and *time frame* available for completion of the project/task as a whole is also determined.
- Draw the relationship diagram (i.e. *precedence diagram*); critical activities to be performed before carrying out the subsequent operations.

- Identify the activities to be performed, serially and in parallel (i.e. *sequentially and concurrently*). Also estimate the *time frame relationships between the various activities.*
- Establish the CPM (critical path to be followed) (Fig. 10.2).
- Carry-out the PERT (Programme evaluation and review technique)

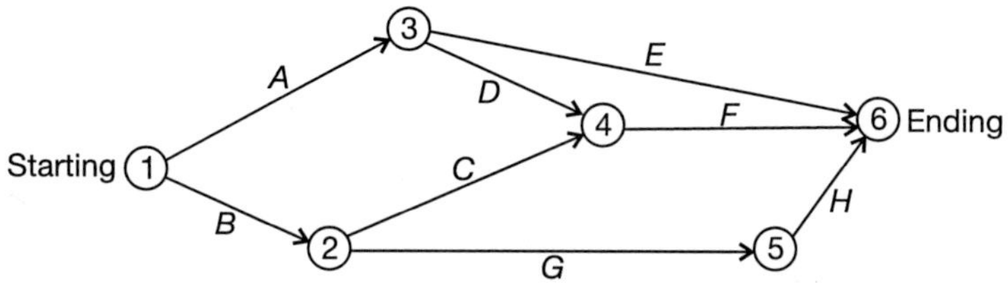

Fig. 10.2: Critical Path Network diagram

Circles; Numbers indicate events (i.e. 1, 2, 3, etc.). Arrows and letters indicate activities (i.e. A, B, C, etc.).

For example, Event 3 is completed after the activity A. Arrows originating at a circle (event) indicate activities that can begin only after all the activities (arrows) terminating at that circle have been completed. Thus precedents for various activities are indicated (known as precedence diagram).

There will be one path through the network that would be *longer than all of the other paths*. Such a path is called "Critical Path" because if any job or activity in that path is delayed, there will be a corresponding delay in the completion time for the entire project (*Time estimates are usually expressed in weeks*). Qualified and experienced managers/engineers will be able to estimate the realistic time required for each activity.

> **CPM**
>
> *Critical Path Method* was developed by DuPont and Remington Rand, around 1960, to plan the construction of a plant (Civil Engineering Project).

CPM can provide management with a no. of alternative schedules. Total project time may vary and so the cost. Computers and softwares enable the calculations to be made easier and quicker. Two types of calculations can be made.

(1) Minimum cost estimate and its time.

(2) Minimum time estimate and its cost.

Based on this management can take a decision for executing the project.

Both CPM and PERT are computer-based scheduling systems for planning and executing the project.

PERT

Programme Evaluation Review Technique was developed in 1958 by US Navy to accelerate the Polaris missile and submarine project; through the use of PERT, they were able to complete the project two years ahead of schedule with a savings of million of Dollars.

CPM is a deterministic system that uses the best estimate of the most likely time. PERT in a Probabilistic system that estimates the time (with a probability attached to the estimated time).

The project manager can then determine the *optimistic time estimate*. With the help of project management softwares *optimum man power requirements* can be calculated. Also, *Time Compression techniques* can be applied to speed up and complete the project.

Precedence Diagram

It is a graphical representation showing that a particular operation or activity is to be carried out (precedence) *before* a subsequent operation for completion of a project/ task. These are referred to as *technological sequencing requirements* (Fig. 10.2) or *Precedence Constraints*.

Examples:
- Drilling of a hole to be done *before*, tapping (thread cutting).
- Washer is placed over the bolt *before* the nut is turned/tightened.
- Switch is mounted to the motor bracket *before* the cover of the appliance is fixed.

Bottleneck Technology Development

Bottlenecks are obstacles which affects or hinders the continuous working of the production line/automation line. Bottlenecks results in less *throughput* of the production line (*Throughput* refers to the output of the line expressed in no. of units produced per hour or per day). Bottlenecks result in accumulation of partly finished components between workstations.

Bottleneck Technology Development involves an understanding of the types of bottlenecks that occur, reasons for it and how to tackle them by developing systems and procedures to minimize them.

Bottleneck machine: A large no. of parts goes through this machine. If this machine fails or not properly maintained, piling up of components will occur. This machine is considered as a bottleneck machine. Sometimes, it is also referred to as *key machine*, which indicates certain *critical operations* are carried out on this machine, as components flow in a manufacturing line.

The capacity of this machine operation can be increased by installing an additional machine and engaging another operator to prevent the piling up of the components. Extra shift for this machine operation can also be considered.

Type of Bottleneck (which obstructs the progress)	Reasons for it, and how to tackle them?
• Stoppage of production	• Shortage of raw material. • Store adequate buffer-stock for timely feeding in to the production/assembly line.
• Frequent breakdown of the line, machines or work stations.	• Machines/equipment not properly maintained. • Develop proper maintenance procedures (by keeping a stock of critical parts for quick replacement). • Training of maintenance staff.

Type of Bottleneck (which obstructs the progress)	Reasons for it, and how to tackle them?
• Building up (Piling) of semi-finished components before a work centre/work station.	• Line balancing is not properly carried out. • Cycle time of this operation is more than that of the average cycle time for earlier operations. • Adjust the working parameters of this operation, by optimizing it so that cycle time is reduced.
• Capacity and no. of *key machines* are not adequate (key machine is wherein a critical operation is carried out).	• Identify the key machine. • Install one more key machine to increase the capacity of this workstation.

Evaluative Questions

1. Explain how a *new product development* can be considered as a project and how can it be planned and managed.
2. What are the *different stages* involved in project management?
3. Explain the meaning of the term "Precedence Constraints". Give some examples from manufacturing technology areas.
4. What is meant by "Bottleneck machine" in the context of an assembly line?
5. What is meant by "cycle time"?
   (Cycle time is defined as the ideal or theoretical time interval between parts coming off the production line/ assembly line.) The required production rate is 60 units/h. At a line efficiency of 100%, calculate the cycle time.
6. What is "Bottleneck Technology"?
   What is the impact of bottleneck in an automation line/ production line?
7. What is "CPM" and "PERT" in the context of project management?
8. Name of two computer-based scheduling systems.
9. How project management softwares help the project manager in planning and executing the project.

Glossary of Terms Related to Concurrent Engineering

Advanced Manufacturing Technology (AMT)

AMT refers to the Computer Integration of MRP (Manufacturing Resource Planning), CAPP (Computer-aided Process Planning), CAM (Computer-aided Manufacturing) including Robots, FMS (Flexible Manufacturing System), CNC (Computer Numerical Control)/DNC (Direct Numerical Control) and CADE (Computer-aided Design and Engineering), and CE (Concurrent Engineering).

Evolution of manufacturing and gradual development of manufacturing technology from the past to the present is shown in Fig. 1.

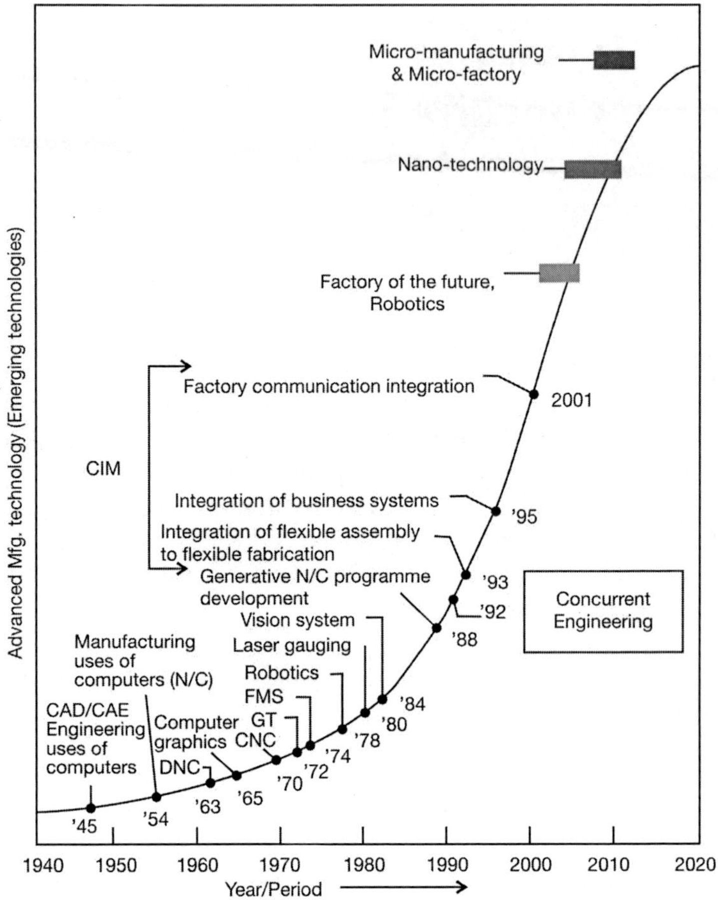

Fig. 1: Global developments in advanced manufacturing technology (Emerging Technologies)

DNC : Direct Numerical Control
CNC : Computer Numerical Control
GT : Group Technology
FMS : Flexible Manufacturing System
CIM : Computer Integrated Manufacturing
CAD : Computer Aided Design
CAE : Computer Aided Engineering
CE : Concurrent Engineering

New Paradigms of AMT (Advanced Manufacturing Technology)

Effective introduction of Advanced Manufacturing Technologies has a pervasive effect on the organization's culture and practices. Existing paradigms and emerging (new) paradigms are:

Organizational Aspect	Traditional Practice (old paradigm)	Shift to Practice Compatible with AMT (new paradigm)
Authority	Based on position	Based on knowledge
Decision-making power	Located close to the top	Located close to the required action
Employee contributions	Limited knowledge and skill	Enhanced knowledge and skill
Information	Closely controlled	Shared widely and used through a no. of media
Rewards	Rewards individual performance	Rewards teamwork and collaboration
Status	Highlights differences in dress, and company	Such differences are reduced
Supervision	Views supervisor as boss	Views supervisor as a resource and facilitator

Agile Manufacturing

Agile manufacturing, is an evolving top-down enterprise-wide effort that supports the concept of lesser "time-to-market" and improving the competitiveness. An agile enterprise could reconfigure their operations, processes and business relationships swiftly considering all market forces, and acting in an environment of continuous and unpredictable change.

Agile manufacturing aims to make company's production system flexible enough to make design changes and design improvements on a continuous basis on the feedback from the customers in a cost-effective manner and in a short time within the overall framework of the company.

Artificial Intelligence (AI)

Acting or behaving intelligently, i.e. acting or behaving like an intelligent person (in contrast to human intelligence or natural intelligence possessed by human beings).

Terms associated with AI:

Data: Collection of unorganized facts, figures, mutually unrelated numbers, characters and symbols.

Information: Organized data based on some criteria or grouping.

Knowledge: Meaningful information obtained by observation, studying and reading.

Intelligence: Ability to understand, discriminate apply and acquisition of new knowledge (application of knowledge gained to new situations).

Artificial Intelligence: Intelligence obtained by observing and reasoning, acting or behaving intelligently, i.e. behaving like an intelligent person (in contrast to human intelligence or natural intelligence possessed by human beings).

Brain Storming

It is a management technique by which a group of people tries to find a solution for a specific problem by gathering a list of ideas spontaneously contributed by its members.

This method is used to generate a large no. of ideas on a specific topic in lesser time by involving many people. The suitability of ideas can be explored later.

Concurrent Engineering

Concurrent Engineering is used to improve the competitiveness in the manufacturing sector. It is a management and engineering philosophy used to improve quality, decrease costs and reduce the lead time to develop new products. A major goal of concurrent engineering is to shift the focus to overall product optimization rather than the optimization of individual department performance. Overall product optimization is an integrated optimization of design, materials, processing, assembly, sales and marketing, and servicing (and maintenance). It is not a specific tool, architecture, technique or process: it is a management and engineering philosophy/concept which must be understood by the top management and implemented in the company across all the functional groups.

Conceptual Design

This the first level when the product is designed. Based on the ideation, only the functional requirements are considered by the designer using technique such as feature-based design and/or solid modelling. Rough magnitude of estimate of time and cost can be obtained at this stage.

Preliminary design and detailed design are next two stages in the design process. Geometry of parts, limits, fits and tolerances and details of materials are determined in these stages.

Computer Aided Process Planning

Process plan: The detailed instructions for making a product. This includes operations to be carried out, machine tools to be used, speed, feed, depth of cut, tolerances, dimensions to be maintained, time standards (cycle time) and inspection procedures and inspection gauges to be used.

Computer Aided Process Planning (CAPP): An interactive computer system that automates some of the steps involved in preparing a process plan.

Computer Aided Engineering (CAE)

CAE is a technology concerned with the use of computers and relevant softwares to analyze CAD (Computer Aided Design) geometry and models allowing designer to simulate and study how the product will actually perform or behave during its service/loading conditions, and making suitable design modifications. Flow analysis of Fluids, Response of electrical and electronic circuits and simulation of complex systems can be studied.

Computer Integrated Manufacturing (CIM)

CIM is a representation of all integration technologies such as CAD, CAM, Concurrent Engineering and Product Data Management. CIM integrates all manufacturing functions and business plans including sales and distribution for successful running of an enterprise. All manufacturing related activities are carried out using computer systems.

Computer Aided Manufacturing (CAM)

The effective utilization of computer technology in the control, operations, and management of the manufacturing facility through either direct or indirect computer in-faced with the physical and human resources of the company (includes CNC programming and CAPP).

Computer Aided Design (CAD)

The use of computer graphics to assist in the design of parts for manufacture and to provide geometric data to manufacturing processes.

Changes in the design can be rapidly implemented on a CAM system.

CAD enables improved quality of designs and better productivity of the designers, and reduced operating expenses.

CIM Implementation

Systems such as CAE, CAD/CAM, CAPP, MRP II and NC (CNC/DNC) were developed to enhance the manufacturing productivity in organizations; and they have been progressively automated. But, they were not designed and implemented for integration. They functioned as "Islands of automation" (Fig. 2).

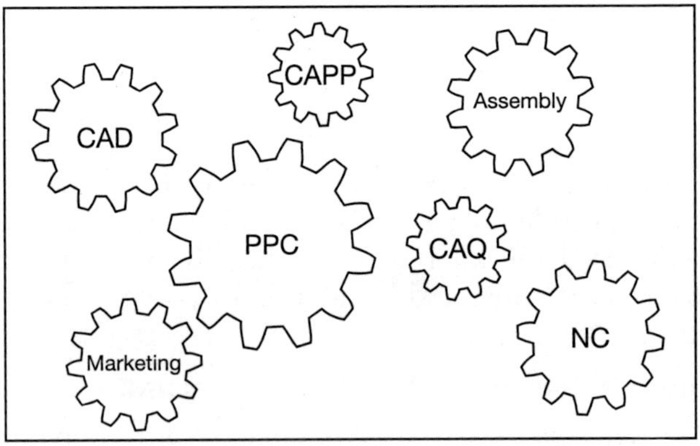

Fig. 2: CIM potential lies in integrating the isolated islands of automation, such as CAD, CAPP, PPC, CAQ and NC (Computer Aided Design, Computer Aided Process Planning, Production Planning and Control, Computer Aided Quality, Numerical Control, Assembly, Marketing, etc.) through a common database

CIM (Computer-integrated Manufacturing) and its potential lies in the successful integration of these "islands of automation" through a common database and relevant softwares by using LAN (Local Area Network) and connectivity for optimum execution of orders (Fig. 3). Many technologies are involved in CIM (Fig. 4). Database required for CIM implementation is shown in Fig. 5.

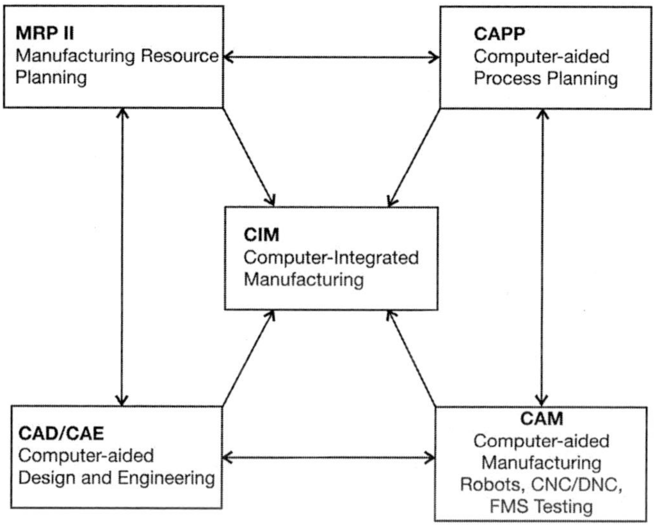

Fig. 3: Integration and connectivity of 4 types of AMT (Advanced Manufacturing Technologies) results in successful implementation of a CIM system.

MRP: Material Requirements Planning (for use of *materials only*)

MRP II: Manufacturing Resources Planning (all *resources*)

CNC/DNC: Computer Numerical Control, Direct Numerical Control

FMS: Flexible Manufacturing System

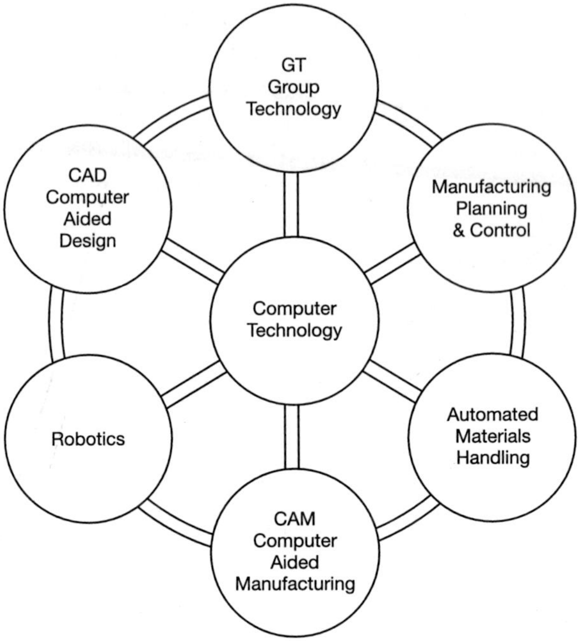

Fig. 4: Many Technologies are involved in CIM. Figure shows the relationships that exist between the major technologies and how *computers serve in the integration role*

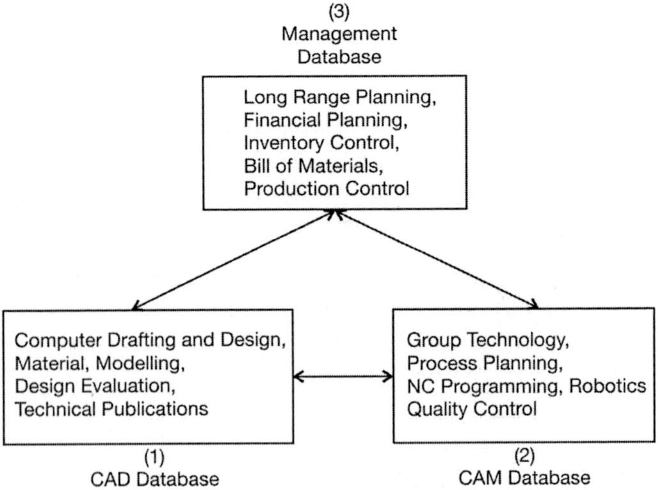

Fig. 5: Three Data bases required for successful CIM implementation. This calls for reliable information storage, retrieval, and information flow

CIM Enterprise model used by researche, and manufacturing engineers is shown in Fig. 6.

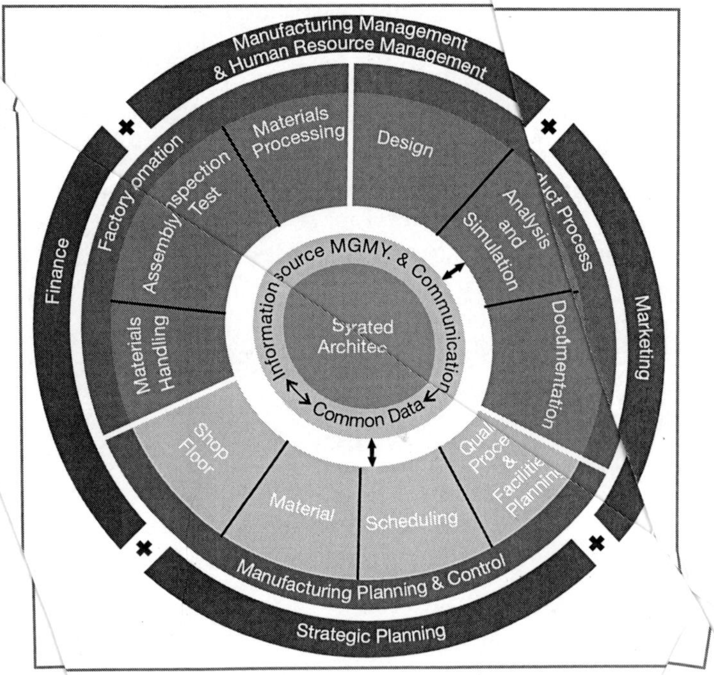

ig. 6: CIM Enterprise Wheel model extensively used
researchers and industries

Commo Database

Electr nically stored manufacturing data use to and accessible r various activities in the plant.

CPM

Critical ath Method, a technique for sc ling and monitoring t accomplishment (completion) c complex project (It is a ool used in project management)

Creativity in Lsign

Creativity in roduct design is imagination ciated with the development new and innovative produc gns. Ability of the designer to cnceive different alternative designs is to be cultivated amongsproduct designers.

Creativity in esign is also associated with designing new products with athetic sense, suitable colour combinations, shape and type finish given to the product. Artistic bent of mind and sense aesthetics are also distinct features of crea...ve product designs and design engineers.

There are techniques by which product des...rs and industrial degners can develop their creativ...ility. These techniques e brain storming, morpholo...al analysis and Edward DeBono's Lateral thinking (of...f-box thinking, i.e. unconventinal and not heard of ...f thinking). Individual and grou efforts will enable ...design department to come out with cceptable and pr...ble design, i.e. design which will result in ess cost.

Desig for Man...cturing (DFM)

It the ...ess of proactively designing products to optimize all the ...ufacturing functions such as fabrication, assembly an...ting. DFM assures quality, reliability, safety and customer sati...tion.

...gn for Assembly (DFA)

DFA takes into consideration how the parts manufactured can be assembled with ease and quickly. Fits, limits and tolerances are suitaly incorporated in the design of parts. In DFA fool-proof orientatn of individual parts is also ensured during assembly by proving suitable matching features, such as chafer, taper and conurs. Also design is ensured so that no damge occurs to parts sembled.

DFM (Design for Manufacturing and Assembly) together takes int nsiderations both manufacturing and ssembly in an integrate anner. Modular design of parts and ub-assemblies are basec the concepts of DFMA.

FMS

FMS efined as "A system dealing with distributed data procg and automated material flo using computer controlled machines, assembly, cells, industrial robots, inspection equipment, together with computer integrated materials handling and storage systems."

Factory of the Future (FoF)

Factory of the future has *four distinct principal characteristics*:

(1) It will be *flexible*,

(2) It will be *automated*,

(3) It will be *robotic*, and

(4) It will be *computer integrated.*

(1) *Flexible* means that the factory of the future will be able to produce a variety of parts/products economically and quickly, capable of changing from part/product to another by means of computer controlled, multi-functional systems for materials processing, materials handling without being limited by part/product dependent tools or part-dependent computer programming. Thus factory of the future will be able to produce goods cost effectively in batches as small as one (i.e. Lot size 1).

(2) *Automated* means that the factory of the future, for a given material and design specifications as input, will be able to produce finished product as output without direct intervention of people in the production processes. Thus, people will design, build and maintain the factory of the future, but they will not, in fact, operate it.

(3) *Robotic* means that the factory of the future will be able to operate autonomously, i.e. as a system which receives sensory input from its environment, compares the input to the expectations desired, chooses the best action among a variety of possible actions (options) based on the results of that comparison, and acts through manipulators in order to carry out the assigned task. Thus, the factory of the future will employ robots extensively; in addition it will be a robotic system by itself as well.

(4) *Computer integrated* means that within factory of the future, all materials, processes, products, managerial and business functions will be represented in computerized data/information

that moves across freely (information flow*) between the various functional units and is used throughout the manufacturing cycle of the product. Thus, in the factory of the future, shared data/information will be integrated with all the sub-components and sub-functions of the computer integrated manufacturing enterprise. Figure 7 shows a pathway to the factory of the future.

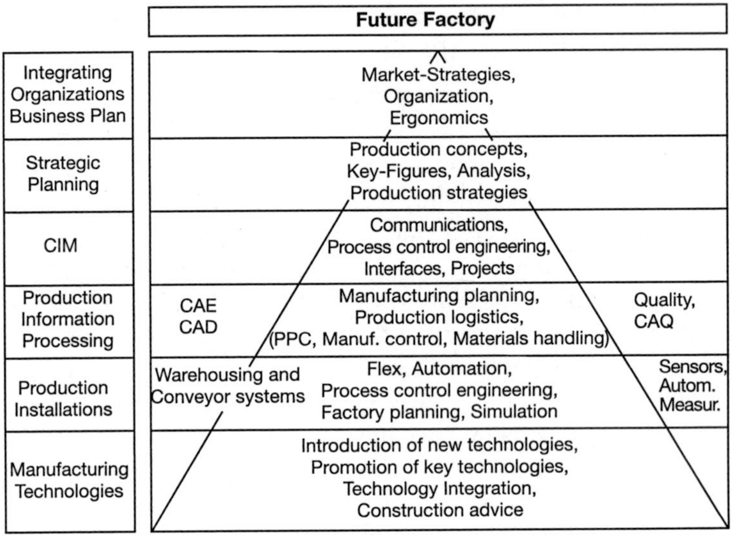

Fig. 7: Pathway to the Factory of the Future (through continuous improvement and automation possibilities)
CAQ: Computer-aided Quality (Control) and
Automated Measurements, (CAI: Computer-aided Inspection)

Flexibility

The term "flexibility" is often used in production systems such as FMC (Flexible Manufacturing Cell) and FMS (Flexible Manufacturing System).

Flexibility is a measure of indicating the level of flexibility (i.e. the degree of flexibility) available in the production system.

* Treatment of information as a factor of production.

Flexibility is defined as *the ratio* between no. of variants/ variations in the design of the products that can be produced to the average batch size in the production system.

$$\text{Flexibility} = \frac{\text{No. of variants produced in the system}}{\text{Average batch size in the system}}$$

If the ratio is greater than 1, larger is the system flexibility (i.e. highly flexible). If the ratio is less than 1, the system is referred to as with low flexibility.

There is no system with a ratio (Quotient) that is approximately one.

Group Technology

The mass production industries are characterized by the manufacturing of large number of *components of the same size and configuration*, and high productivity is attained by the high degree of automation, special purpose machinery, and flow production methods. Such production system and equipment needs high capital investment and they are not flexible in operation.

The majority part of the industry, however, concerned with production in medium and small batches. Small batches and large variety of components which need to be produced have prevented the extensive use of automation and production is carried out using general purpose machinery with high degree of flexibility.

Group technology is an approach or technique to batch manufacture which aims to increase productivity by following techniques as flow-line and low-cost automation methods of production.

Group technology *identifies and brings together related or similar components* in a production process in order to take advantage of their similarities by making use of inherent economics of flow production methods; i.e. advantages of flow-line production (which is normally associated with automated manufacturing systems) are realized by group technology.

Part families (i.e. groups of parts formed based on their similarities) *are taken through the manufacturing processes as a group* (Figs. 8 and 9).

Fig. 8: A family of similar components (Flange type of components) constitute a Group

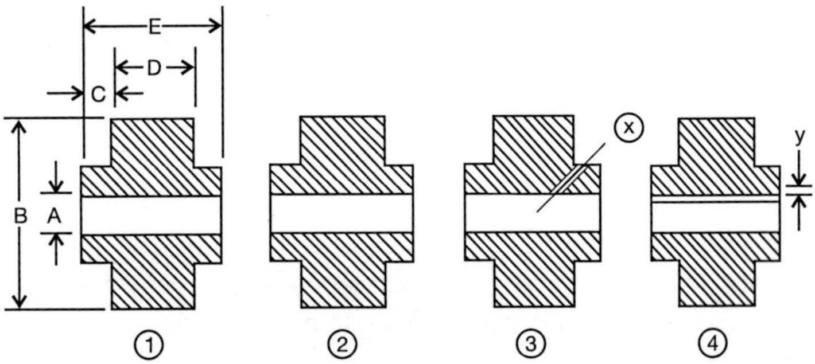

Fig. 9: Parts family having similar processing operations Turning Facing, Drilling

How is a group formed? (criteria for forming of families)

A family of similar components constitute a group (Overall component shape and size is not important).

Family of similar parts must have two basic characteristics in common.

(1) They must have the same holding or chucking requirement.
(2) They need the same or similar surfaces to be machined in terms of spatial location and size.

All Flange type of components constitute a group (as in Fig. 10).

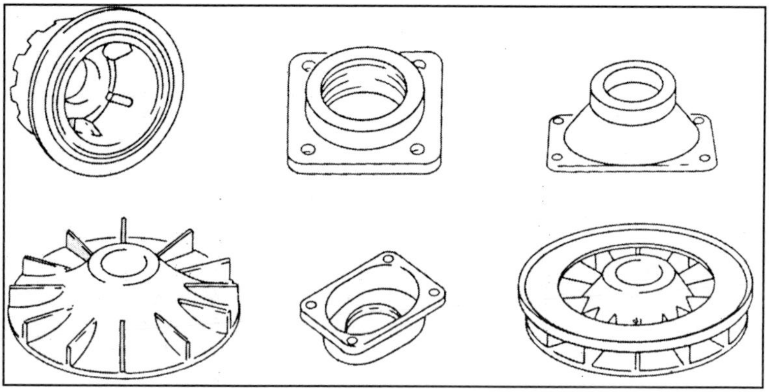

Fig. 10: Parts family have similar shapes: Flange and rib shaped parts

This technique was first conceived by R.E. Flanders, Transaction of ASME, USA, (1925); Vol. 46, pp. 691-738, and implemented in UK industries successfully (1973).**

Four stages are involved in implementing group technology (Fig. 11).

> Group Technology application in manufacturing was first developed by S.P. Mitrofanov, a Russian Engineer, 1959.
>
> "Scientific Principles of Group Technology"

** *Group Technology*, C.C. Gallagher and W.A. Knight, UK, London.

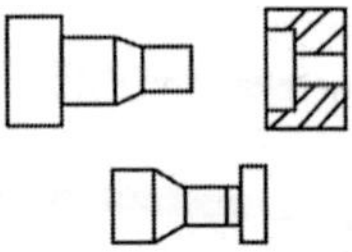	1. Stage: Part Family Parts with similar manufacturing requirements (grouping parts)
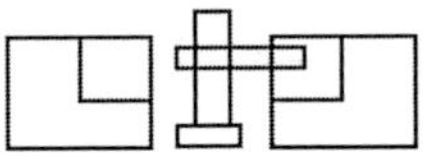	2. Stage: Manufacturing Facility Equipment needed for completely manufacturing a part family (grouping machinery)
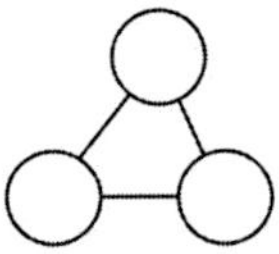	3. Stage: Working Group Equally skilled workers cooperating to completely manufacture a part family by appropriate equipment (grouping personnel)
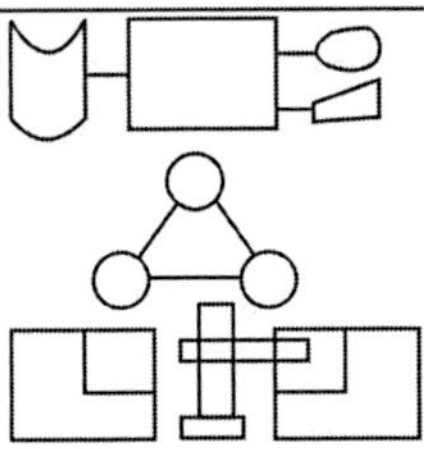	4. Stage: Production Integration of design, planning and controling tasks for complete production of a part family (organizational grouping)

Principles of group technology

Fig. 11: Group Technology (GT) is implemented in four stages according to their (company) organizational features.

Stage 1. Formation of part families depending on their similarities
Stage 2. Grouping of machinery for manufacturing
Stage 3. Grouping of personnel according to their skills
Stage 4. Production of complete part families (organizational planning)

Gantt Chart: A bar chart used to monitor the progress of a project.

Industrial Robots

The word Robot was coined, in 1920, by Czech engineer in his paper: Rossum's Universal Robots, it is derived from Czech word "*Robota*", meaning "*worker*".

An industrial robot has been defined as a reprogrammable multi-functional manipulator, designed to move *materials, parts, tools, or other devices* by means of variable programmed motions and to perform a variety of other tasks. The term robot also include manipulators that are activated directly by an operator. *The tool or device attached to the robot wrist is called an "end-effector" which carries out the operation.*

According to *ISO* (*International Organization for Standardization*), an industrial robot has been defined as follows:

"A machine formed by a mechanism including several degrees of freedom, often having the appearance of one or several arms ending in a wrist capable of holding a part, tool or device." In particular, its control unit (computer) uses a memorizing device and it may sometimes use sensing devices to take into account environmental factors (such as, hot, hazardous, toxic).

These multi-purpose machines are designed to carry-out a repetitive function and can be adapted (programmed) to carry-out other operations. Picture of an 6-axis *industrial robot* is shown in Fig. 12.

Industrial robots are (i) hydraulically or pneumatically operated, (ii) servo-operated, and (iii) computer controlled.

The six Rotary axis movements are:

(1) Arm Sweep

(2) Shoulder Swivel

(3) Elbow Extension

In addition, the orientations of wrist are:

(4) Yaw

(5) Pitch

(6) Roll

It has jointed configuration. Arm can reach: 3m vertically upward and extended out horizontally upto 2.5m while travelling through 240° of arm sweep. This works out a total working volume of about 30 litres, i.e. any spot in the working volume (called work envelope of the robot) can be reached by the wrist.

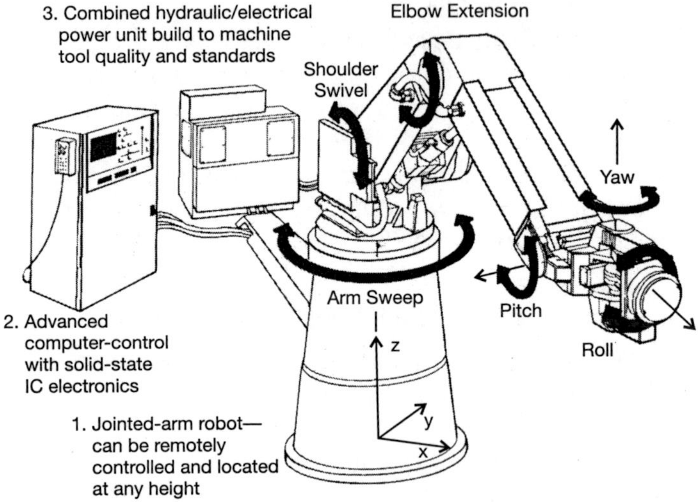

Fig. 12: Six-axis industrial robot.
- Repeatability can be upto ± 0.2 mm
- Pay load at the wrist: 10 kg

The Tool Center Point (TCP) is a point that lies along the roll axis at a specified distance from the roll face plate. All orientations of the wrist axes takes place about the Tool Center Point, and this is considered when programming a robot to reach a specified location for any operation.

The travel speed of a standard arm: 1m/sec.

A typical control unit has 23k byte mini-computer. The locations of 1000 points (TCP) can be pre-programmed and stored in the memory.

Coordinate systems of a robot configuration is shown in Fig. 13.

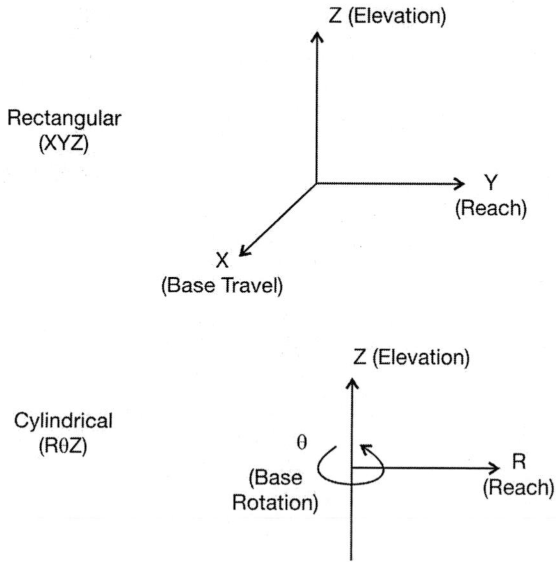

Fig. 13: Coordinate systems used in Robotics; Rectangular coordinates and cylindrical coordinates

Pick-and-place robot is shown in Fig. 14.

Fig. 14: A Pick-and-Place Robot

An industrial robot in action is shown in Fig. 15.

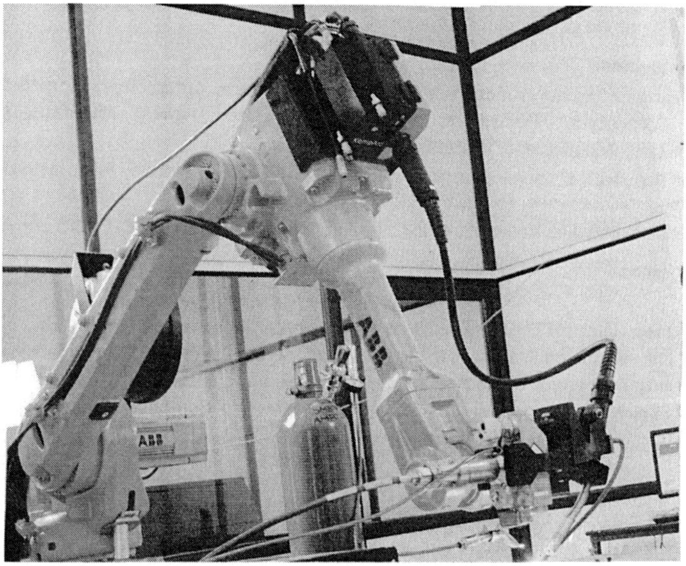

Fig. 15: A Robot in action

The use of computer data bases to track inventory levels at any instants warehousing/stores information, despatches to production department and stock received from suppliers/ vendors.

Innovation

(1) *Process Innovation*: Process Innovation involves incremental production process improvements leading to an efficient production system specially designed to produce highly standardized products. New machinery and equipment may need to be installed for process innovation.

(2) *Product Innovation*: Traditionally product innovation is referred to as R&D (Research and Development). Product innovation is creation and evaluation of new products and services. Product characteristics undergo radical changes. New products require reorientation of company's goals and production facilities. "High Technology" industries are born out of product innovation.

Examples of "*High Technology Industries*" are: Defence, aircraft and aerospace, communication equipment, electrical and electronics equipment, plastics, resins, and industrial adhesives. Relatively high R&D expenditures are involved.

"*Low Technology industries*" are mining, steel making, pulp and paper manufacturing and industrial chemicals. Relatively low R&D expenditures are involved.

Just-in-Time. It is a type of production system wherein large buffer inventory is not kept. In fact, the *batch size* in an ideal JIT concept *is only one.*

Kaizen

Kaizen a Japanese concept for *continuous improvement of products, processes and quality*. Quality controls are identified, adopted and implemented with TEI (Total Employee Involvement). Groups of operators, managers and engineers (members of quality circle) directly concerned with the production operations are involved in the implementation of Kaizen plans.

Lean Production

Lean production refers to producing more with less. This concept was first introduced by Womack, Jones and Roos ("The machine that changed the world", 1990). Lean production is the result of use of several newer developments and technologies that are being adopted by manufacturing industries. The basic underlying principle is waste elimination in the use of all resources including people, effort and time.

Morphological Analysis

A method of identifying different sets or combinations of product characteristics and examining the relationships among them. The technique is used to generate new product ideas.

Mind Mapping: It is a two-dimensional graphical method (or representation) of organizing concepts, theories and thoughts. Here understanding is schematically represented by creating a hierarchy of ideas/concepts (Trees) linked through branches of sub-concepts. Inter-relationships are indicated by showing cross-links or additional branches. Integrated knowledge or global perspective of a topic can be shown for easier understanding.

Mind mapping is also known as *Concept Mapping*, has been formulated by a Psychologist, Tony Buzan, 1970. Concept mapping is also referred to as "*Knowledge Map*"/"*Knowledge Mapping*".

Machine Tool Capability. Capability evaluations of equipment used for manufacturing in a shop floor. Machine tool capability takes into consideration the capacity of the machine tool (size and power rating), ability to meet close dimensional tolerances and surface finish requirements.

Network Analysis

Network analysis use a technique (such as PERT) for planning and control. Involves identifying activities and events that must take place and their timing in relationship to other activities necessary for the completion of the project/product by the target date.

Network Diagram: A diagram used in network analysis, representing all the activities and events that must be accomplished. It is a flow chart showing the sequence, precedence, if any, and the interrelationship of the planned activities, tasks and events.

New Product Development: Development of a new product which is new to the company producing it and to the markets in which it is to be sold. Sometimes, it also includes products which have been modified from others in terms of changes in some physical characteristics, new features, brand name, and packaging.

New Product Development Department: A functional entity in a company whose specific responsibility is the generation and scrutiny of new product ideas. Product development activities are directed and coordinated by this group. The Dept. is headed by an executive officer/new product manager. (Product modification involving changing or adapting a product to meet the demands and needs of a target market is also considered as the responsibility of this Dept. or Team.)

Problem-Solving Techniques

There are many problem-solving techniques, which are useful in new product design and development. These are:

- *Affinity Diagrams*: Organizing the ideas/requirements into common and related themes.
- *Matrix Diagram*: Understanding different aspects, risks and their implications.
- *Appreciating Objectively*: Extracting maximum information from facts and figures and understanding them.
- *Appreciative Enquiry*: Looking at what is going on and solving the problem.
- *Brain Storming*: Generating a large no. of ideas from a group of people to solve a problem.
- *Cause and Effect Diagram (Fish-bone Diagram)*: Examining the problem, identifying the possible causes and factors that influence the end-effect.
- *Core Competency Analysis*: Identifying core competency areas and skills, staying ahead and going ahead.
- *Critical Success Factors*: Identifying critical factors that really matters for success.
- *Five W's Technique*: What, why, who, where and when: using questioning techniques, quickly getting to the root cause and solving the problem. "How" can also be added as the questing technique.
- *Polya's Principle* (Polya, G., Princeton University, USA, 1988): Polya breaks the problem-solving process down into four main parts: understanding the problem; devising a plan, carrying out the plan, and looking back.
- *Flow Charts*: Understanding how the materials flow and information flow occurs, and process work.
- *Analyzing the Crisis*: (Accident Analysis in respect of safety management in industries and traffic management/ road safety.) Surviving the crisis and that comes with the growth, in industries and society.

- *Lateral Thinking* (Edward De Bono's Lateral Thinking): Also known as Out-of-Box thinking, changing/inversing the concepts and perception.
- *Working in Harmony*: Making sure that all groups of an organization work in harmony and unison.
- *PEST Analysis* (Political, Economical, Socio-cultural and Technological analysis): Understanding the environment and the whole picture, scenario.
- *Understanding the Power Center*: Understanding where the power lies, and working accordingly.
- *Risk Analysis*.
- *System Diagram/System Model*: Understanding the system, sub-system, objectives, input and process variables and their influence on outcome; and feedback and control.
- *Root Cause Analysis*: Identifying various factors and causes for problem-solving.
- *SWOT* (Strength, weakness, opportunities and threats): Analysis, used in business situations.
- *USP* (Unique selling proposition): Analysis, identifying unique features and evolving strategies based on it.

Process Plan

The detailed instructions for making a part or a product. This includes operations to be carried out, their sequence, machine tools to be used, machining conditions to be used (such as cutting speed, feed rate and depth of cut), dimensions to be achieved with their tolerances specified, special tooling devices, such as jigs, fixtures and gauges to be used and cycle time, inspection procedures and any heat treatment required. The process plan is also referred to as "*Route Sheet*" (since this indicate the route or sequence to be followed). Process plan is prepared by methods engineer (or) by process planning department.

In essence, the preparation of process plan involves five basic steps.

(1) The required operations and their sequence must be determined by studying the part design or component

drawing. The sequencing of operations is known as *technological sequencing*.

(2) The machines required for each operation must be determined. The selection of machine tool depends on part size, power, torque, surface finish and accuracy required.

(3) The required tools for each operation/machine must be determined. Gauges required are also identified.

(4) Optimum machining parameters are determined, depending on production rate, finish required and type of tool used.

(5) An estimation of machining time or cycle time is done.

A *chart* (*known as Route sheet*) is prepared in a Tabular form following the standard practice adopted by the company.

CAPP (*Computer-aided Process Planning*) involves the use of computers to automate this process.

Process Planning

The preparation of a set of instructions that describe how to fabricate a part or build an assembly which will satisfy the engineering design specifications.

Computer-aided Process Planning: There are two types of CAPP:

(1) *Variant Process Planning*: This is also called as retrieval method. This involves a library of standard process plans; interactive editing programmes; and storage searching and retrieval, and documentation capabilities. The process plan for a new part is created by *retrieving* and *modifying* the standard process plan for a similar part (from a given part family). The term *Variant* indicates the process plan is slightly at variance from the process plan of a similar part.

(2) *Generative Process Planning*: A *process planning system* consisting of a database and decision logic, that will *automatically generate a process plan* from graphical and textual specifications of the part.

CAPP (Computer-aided Process Planning)
It is an *interactive computer system* that automates some of the steps involved in preparing a process plan.
The process planner (process engineer) checks, validates and gives final touches to the process generated by the computer system.

The generative process planning gives the *optimal process plan* for manufacturing particular product/component.

AI (Artificial Intelligence) and ES (Expert Systems) enhance the effectiveness of computer-aided process planning.

Production Managers

The main concerns of Production Managers are (in the order of priority):

1. Rising overhead cost
2. Producing to high quality standards
3. Introducing new products on schedule
4. Rising material cost
5. Availability of qualified engineers and staff
6. Delivery on time
7. Poor sales forecasts
8. Making new process technology
9. Falling behind in process technology
10. Rising inventories
11. Excessive lead times
12. Low productivity in the field of indirect costs

Rapid Product Development

Rapid product development is a manufacturing philosophy adopted to shorten or reduce the product development time. Rapid product development is the synergic (i.e. simultaneous) integration of various time compression technologies which may be physical in nature or virtual (i.e. computer based). Time compression can be achieved by carrying out the design, manufacturing, assembly and testing activities in the virtual world, i.e. on computer systems, as far as possible. When this is not possible, designers and manufacturing engineers can resort to physical world (i.e. actual fabrication and manufacturing) where all efforts are made to do them as fast as possible, and prototypes are developed for visualization and for rapid evaluation of its manufacturability and its design effectiveness and for concept validation.

Rapid Prototyping

Prototypes are made quickly (rapidly) using special equipment such as Stereo Lithography (STL) Apparatus, *Selective Laser Sintering* (SLS) and *3-D printing machines*. In Rapid prototyping materials such as wax and polystyrene are melted and deposited layer by layer, the material solidifies and the component or model is built-up.

Rapid prototype helps in making the models/prototypes quickly for visualization (i.e. to know how the component will look like) and for making any design modifications before actual production.

Tooling elements (dies and moulds) can also be made quickly which is called Rapid tooling.

Virtual Prototyping

Virtual Prototyping is a software based methodology of producing a prototype using advanced 3D graphics and virtual reality environments.

This enables designers to design, develop and examine a part on a computer screen. This technology uses a CAD package to shape a part so that it can be evaluated as it is drawn.

Virtual prototyping systems can be used to render the intricate part details on a screen.

Virtual reality equipment uses a specialized design "head gear" and "gloves" with appropriate sensors to observe a computer generated prototype on a virtual environment. Advantage of this is quick rendering of the parts instantaneously.

However, many manufacturing and design engineers prefer to see a physical model (mechanical system) to evaluate, rather than on a video screen rendering.

A high-profile application of this has been the design, development and shaping of Boeing 777 aircraft parts where fits and interferences were evaluated on a CAD system and difficulties corrected before manufacturing first production model.

Reverse Engineering

Reverse engineering is used to create drawings, designs and geometric models (CAD models) from actual physical objects already available. Reverse engineering is adopted in circumstances where components, especially of old and worn-out equipment, actually exist, but their original drawings are no longer available, but parts need to be produced again.

In reverse engineering, the component or object is placed on a CMM (Coordinate Measuring Machine), a stylus or a touch probe, traces the contours and features (sizes, dimensions and forms), and recorded on a computer system and retrieved later. The designs, drawings and geometric models created help in manufacturing the component.

Manufacturing of artificial limbs in medical prosthesis is an application of reverse engineering.

Standardization

Establishment of basic limits or standards in the form of specifications to which a product should conform. Also, involves use of standard components for an assembly, eliminating a variety of machine elements in an assembly.

Specifications: These are engineering and technical information presented in a document giving description of items/ details such as dimensions, tolerances, fits, electrical ratings and other information useful for buyers and maintenance staff. This is particularly important for industrial goods and equipment.

Throughput

Generally refers to output of an automatic production line, expressed in units produced per hour or units produced per day. Reduction in throughput is due to bottlenecks in the production line.

Quality Function Deployment (QFD)

Quality Function Deployment is a technique in concurrent engineering used in product planning to build into the product what the customer needs. By employing QFD, the design is finalized before the production starts.

Value Analysis

Value Analysis is defined as "A systematic interdisciplinary examination of factors affecting the cost of a product or service in order to devise means of achieving the specified purpose most economically at the required standard of quality and reliability."

The impact of VA (Value Analysis) is in four main areas:

1. Redesign (or products/parts)
2. Choosing alternative materials
3. Elimination of unnecessary features (part details)
4. Changes in procedures of making parts.

The term "Value Analysis" was coined by H.A. Winn, President, GEC (General Electric Company), USA, 1974.

Workstation

A workplace in an assembly line, where manufacturing activity takes place, and the material undergoes the process of becoming a finished part (or product).

University Model Question Paper

Time: 3 Hrs. Max. Marks: 100

Part A (2 marks each × 20: 40 Marks)
(Answer all the Questions)

Choose the correct alternative and write the corresponding letter in the Answer Sheet:

1. The application and use of Jigs and Fixtures in manufacturing operations.
   (a) Increases the production rate
   (b) Widens the technological capabilities of machine tools
   (c) Enables automation fully or partially
   (d) All of the above.
2. The heart of automation technology is
   (a) Safety of the operator
   (b) Robots and robotic devices
   (c) Computers
   (d) Reliable sensors
   (e) Control system employed.
3. A continuous process is
   (a) Used only for simple tasks
   (b) One which is never shut down
   (c) One in which raw materials enter the process and a specific product comes out
   (d) One wherein large no. of robots are used.
4. The important difference between open-loop control system and closed-loop control system is that
   (a) Open-loop systems are less complex
   (b) Open-loop systems are less expensive
   (c) Open-loop systems are less accurate
   (d) In open-loop system no information from the output is fed back to the input.

5. "The products are made in small lots and in large variety." This method of production is called
   (a) Mass production
   (b) Continuous production
   (c) Semi-continuous production
   (d) Batch production.
6. The concept of "ABC" analysis is applied in
   (a) Plant layout and materials handling
   (b) Ergonomics
   (c) Inventory control
   (d) CAPP (Computer-aided Process Planning)
7. Flexible manufacturing (FM) Process Planning
   (a) Automated design
   (b) Tool design and tool making
   (c) Quick and inexpensive product changes
   (d) Large volume of production
8. ________ is concerned with the activities that are involved in the preparation of route sheets. (Route sheet lists the sequence of operations required to produce a given part/component.)
   (a) CAD (b) CAM
   (c) CADD (d) CAPP
9. _________ involves the use of computer systems to plan, manage and control the operations of a manufacturing/production plant through either direct or indirect computer interface with the plant's resources.
   (a) CAD (b) CAM
   (c) CAI (d) CAPP
10. In product data management, product designs are documented and stored in computer systems by means of
    (a) Component drawings
    (b) Specifications
    (c) Bill of materials
    (d) All of the above

11. An automated flow line consists of several ________ linked together by work handling devices that transfer parts between different work stations.
    (a) CNC machines
    (b) Robots
    (c) both CNC machines and Robots.
12. Two basic models used in data base management are _____ and ________.
    (a) Synthetic and analytical
    (b) Vertically integrated structure and horizontally integrated structure
    (c) Relational and hierarchical
    (d) Networks and nodes.
13. Which is the best example of real-time processing?
    (a) Summarizing daily cash transactions
    (b) Calculating weekly inventory consumption
    (c) Printing of monthly pay roll cheques
    (d) Requesting an airline reservation.
14. A MIS (Manufacturing Information System)
    (a) consists primarily of equipment and technology used
    (b) consists of rules, procedures, data, hardware, software and persons
    (c) is an enabling tool for taking correct decisions
    (d) in a tool for processing of data and information.
15. In break-even analysis of manufacturing cost, total cost consists of
    (a) Fixed cost + variable cost
    (b) Fixed cost + variable cost + overhead charges
    (c) Equipment cost + cost of power/energy consumed
    (d) Fixed cost + variable cost + overhead charge + profit
16. Porter's model best describes the
    (a) Financial health of the company
    (b) Manufacturability of a product
    (c) Profitability of the company
    (d) Manufacturing competitiveness of the company.

17. Concurrent Engineering is to be followed by
    (a) Product Design
    (b) Discussion with vendors & suppliers
    (c) Discussion with sales & marketing people
    (d) All of the above
    (e) None of the above
18. CPM and PERT are used in
    (a) Sales forecasting
    (b) Estimation of machining time
    (c) Project management
    (d) The analysis of manufacturing competitiveness
19. At what stage in the product life cycle, new products are designed and developed?
    (a) Growth and development in sales
    (b) Maturity
    (c) Stagnation (or) saturation
    (d) Decline
20. At the break-even point, in cost analysis the
    (a) Total cost is more than the sales revenue
    (b) Total cost is less than the sales revenue
    (c) Fixed cost is equal to variable cost
    (d) Total cost is equal to the sales revenue

Answers

1. (d)	2. (e)	3. (c)	4. (d)	5. (d)	6. (c)
7. (c)	8. (d)	9. (b)	10. (d)	11. (c)	12. (c)
13. (d)	14. (b)	15. (a)	16. (d)	17. (e)	18. (c)
19. (d)	20. (d)				

Part B (12 marks each × 5 : 60 marks)
Answer *any five* Questions only

1. (a) What is meant by concurrent engineering?
   (b) What is its relevance to manufacturing industries?
   (c) Mention *any two* important advantages of concurrent engineering.
2. (a) Is "Simultaneous engineering" and "Concurrent Engineering" are one and the same. Say Yes or No substantiate your answer.
   (b) What are cross-functional teams or groups?
   (c) What do they do?
3. (a) What is "Product Development"?
   (b) State the importance of "Modular design" in the context of manufacturing engineering.
   (c) What is product engineering?
4. (a) What is meant by "inventory"?
   (b) What is the relationship between "JIT system" and inventory?
   (c) What are the different types of inventory?
5. (a) How is Information Technology used in product data management?
   (b) Write a paragraph on AI (Artificial Intelligence) and ES (Expert Systems).
6. (a) What is meant by manufacturing competitiveness?
   (b) Explain "Porter's Model" of competitiveness.
7. (a) What is "Product Life Cycle"?
   (b) What are the distinct stages in product life cycle?
   (c) How is "Product Life cycle" useful in the identification and development of new products?
8. Write short notes on any three of the following:
   (a) Precedence diagram
   (b) CPM (Critical Path Method)
   (c) Computer-aided Process Planning (CAPP)
   (d) Solid Modelling
   (e) Design for Manufacturing
   (f) Economic Lot Size (ELS)

Index

A

Adequate, 8, 105-06

Advanced Manufacturing Technology (AMT), 43, 51, 107, 109, 113

Agile, 38, 109

Air-Traffic Control (ATC), 26

Alpha-Numeric Codes, 93

Application, 26

Artificial Intelligence (AI), 11, 25-26, 28, 110, 133

Automated, 48, 60, 62, 64-67, 117-18, 138

B

Bottleneck Technology Development, 105

Brain storming, 11

Budget, 34, 102

Business functions, 14

C

Collaboration, 8

Collaborative companies, 24

Common Database, 115

Computer Aided Design (CAD), 112

Computer Aided Engineering (CAE), 111

Computer Aided Manufacturing (CAM), 112

Computer Aided Process Planning (CAPP), 25-26

Computer Integrated Manufacturing (CIM), 111, 113

Computer Numerical Control, 107-08, 113

Coordinate Measuring Machine (CMM), 135

Critical dimensions, 49

Critical Path Method (CPM), 11, 103-04, 115

Critical Success Factors, 130

Cycle time, 93, 106

D

Data, 15, 18-22, 25, 28, 110-11, 114

Data Management, 18-22, 28, 111

De Bono, Edward, 11, 116, 131

Decision, 26, 72, 109

Design, 1-2, 5, 8, 14-15, 20, 23, 29, 32-36, 39-40, 42, 46-48, 50, 52-53, 57-60, 62, 70, 72-76, 84, 90, 92, 94-97, 100-02, 107-08, 111-12, 115-16, 140-41

Design for Manufacturing (DFM), 73, 116
Design for manufacturing and assembly (DFMA), 75
Designs and geometric models, 135
Direct Numerical Control, 107-08, 113

E

Elements of Product, 1, 95
Expert System, 11, 25-26, 28, 133, 141

F

Factory of the Future (FoF), 117-18
Finite Element Analysis (FEA), 17-19
Flexibility, 118-19

G

Geometric Modelling, 15-16, 90
Global Outlook, 8
Group Technology, 108, 119, 121-22

I

Improved Design Quality, 6
Industrial Robots, 123
Information Technology, 13-15, 141
Innovation, 31, 127
Inventory Control, 57
Islands of automation, 112
IT enables implementation, 14

J

JIT (Just-in-Time), 51, 54, 128

K

Kaizen, 128

L

Lean production, 128
Local Area Network, 113

M

Machine Tool Capability, 129
Manufacturing Lead Time (MLT), 7
Master File, 19
Matrix Diagram, 130
Micro-wave Ovens, 26
Modular Design, 57-59, 74, 141
Moment of Inertia, 18
Morphological Analysis, 11, 128
Multi National Companies (MNC), 22

N

Network Analysis, 129
Numerical Control (NC), 112

O

Original Equipment Manufacturers (OEM), 39, 51, 54, 57
Optimal Design, 36-37

P

PDM integrates, 20
Physical location, 10
Porter, M.E., 43
Porter's Model, 43, 53, 141
Product Data Management (PDM), 18, 22, 28
Production Managers, 133

Productivity, 21

Product Life Cycle, 1, 20, 29, 31, 40, 70-71, 95, 141

Product Realization Taxonomy, 96, 98

Programme Evaluation and Review Technique (PERT), 11

Programming and coding, 27

Q

Quality Function Deployment (QFD), 11, 136

Quick design, 14

R

R.E. Flanders, 121

Rapid Product Development, 133

Rapid Prototyping, 134

Reserve Engineering, 135

Robotic, 26, 117

Robotic Technology, 62

Route Sheet, 34, 39, 131-32, 138

S

Selective Laser Sintering (SLS), 134

Software Development Cycle, 27

SWOT, 11, 131

T

Taxonomy, 92, 94

The ratio, 118-19

Time-to-Market, 8, 21, 109

Tolerances, 85

Tool Center Point (TCP), 124

Total Employee Involvement (TEI), 56, 128

Total quantity, 34

U

Uni-directional information, 5

Unique, 23, 38, 131

Unique selling proposition (USP), 23, 38, 131

V

Value Analysis, 11, 136

Variant Process Planning, 132

Virtual Prototyping, 134

Z

Zooming, 15